UNIPA Springer Series

The **UNIPA Springer Series** publishes single and co-authored thematic collected volumes, monographs, handbooks and advanced textbooks on specific issues of particular relevance in six core scientific areas. The issues may be interdisciplinary or within one specific area of interest. Manuscripts are invited for publication in the following fields of study:

1. Clinical Medicine;
2. Biomedical and Life Sciences;
3. Engineering and Physical Sciences;
4. Mathematics, Statistics and Computer Science;
5. Business, Economics and Law;
6. Human, Behavioral and Social Sciences.

Manuscripts submitted to the series are peer reviewed for scientific rigor followed by the usual Springer standards of editing, production, marketing and distribution. The series will allow authors to showcase their research within the context of a dynamic multidisciplinary platform. The series is open to academics from the University of Palermo but also from other universities around the world. Both scientific and teaching contributions are welcome in this series. The editorial products are addressed to researchers and students and will be published in the English language.

The volumes of the series are single-blind peer-reviewed.

Book proposals can be submitted to the *UNIPA Springer Series Technical Secretariat* at unipaspringer@unipa.it
At the following link, you can find some specific information about the submission procedure to the Editorial Board: https://www.unipa.it/strutture/springer/.

Francesca Sunseri

Psychology and Technology in Gilbert Simondon

UNIVERSITÀ
DEGLI STUDI
DI PALERMO

Springer

Francesca Sunseri
Department of Humanities
University of Palermo
Palermo, Italy

ISSN 2366-7516 ISSN 2366-7524 (electronic)
UNIPA Springer Series
ISBN 978-3-032-25739-0 ISBN 978-3-032-25740-6 (eBook)
https://doi.org/10.1007/978-3-032-25740-6

This Springer imprint is published by the registered company Springer Nature Switzerland AG
The registered company address is: Gewerbestrasse 11, 6330 Cham, Switzerland

Foreword—The Other Gaze of the Technical Object

Several years ago, Jean-Hugues Barthélémy—who in recent years has published numerous volumes elaborating his own philosophical perspective (2025, 2022, 2018), matured also in dialogue with Gilbert Simondon, of whom he is without doubt the foremost living exegete (Barthélémy 2005a, 2005b, 2008, 2017)—invited me to write an article on the various historical stages of the European and global reception of the thought of the philosopher from Saint-Étienne (Carrozzini 2018), whose work I have studied since as far back as 2004. That article was recently republished in an updated version, enriched by the new directions that current debates and research into Simondon's philosophy have opened up (Carrozzini 2024). In its revised and corrected form, the essay appeared in 2024 in the online journal «texto! Textes & Cultures». Yet, paradoxically, owing especially to the numerous events organized in celebration of the centenary of Gilbert Simondon's birth, even this short article already feels dated—at least in the judgment of its author. In the brief span of a single year, publications on Simondon and translations of his writings into several languages have grown exponentially, as confirmed by the regularly updated website *gilbert.simondon.fr*, curated by his family, particularly by his daughter Nathalie and his granddaughter Irlande Saurin.

The text by Francesca Sunseri, for which this essay serves as *Foreword*, makes a significant contribution to that acceleration—an acceleration both necessary and long-awaited—of the analytical yet innovative studies of Simondon's thought. Sunseri, along with other scholars equally rigorous in their analyses—for example, Giovanni Menegalle (2024) in the Italian context—offers readers a clear and engaging overview of Simondon's philosophy. She approaches it from a precise interpretative angle: the contribution of the twentieth-century French philosopher to the fields of psychology and scientific psychosociology, both understood by the Author as theoretical frameworks capable of tracing a systemic unity—open-ended and evolving—within Simondon's philosophy.

As Sunseri rightly emphasizes in her book, Simondon's deep interest in psychology—an interest that culminated in his appointments first at the Faculté de Lettres et Sciences humaines of the University of Poitiers (1955–1963) and later at the Sorbonne (1963–1984), where he held the chair in General Psychology—had already been highlighted a few months after his untimely death by his colleague at the Parisian university, Denise Van Caneghem. In 1989, Van Caneghem published a substantial article in the Bulletin de Psychologie entitled *Hommage à Gilbert Simondon* (Van Caneghem 1989; Carrozzini 2015). In the subsequent phases of rediscovery of Simondon's thought—whether in the period stretching, albeit intermittently, from his death to the dawn of the twenty-first century, or in the later phase which, with notable transformations, continues even now—scholars such as the aforementioned Barthélémy, the Belgian philosopher Pascal Chabot (2003; 2013), the Bachelardian philosopher of science Vincent Bontems (2010), the Italian political philosopher Andrea Bardin (2015), and, of course, the late Bernard Stiegler have primarily focused on Simondon's ontology and his reflections on the protean universe of technical objects, often considering their connections, implicit or explicit, to socio-political questions.

In truth—and I say this without false modesty, as Sunseri herself acknowledges—back in 2011 I published, with Mimesis/Centro Internazionale Insubrico, a substantial extract from my doctoral dissertation in historical-philosophical disciplines, entitled *Gilbert Simondon filosofo della mentalité technique* (Carrozzini 2011). The second chapter of that work was devoted to tracing a theoretical thread running through the numerous courses in *Psychologie générale* that Simondon taught between 1963 and 1980, then available. Nevertheless, my explicit programmatic aim, already evident in the title of that publication, was to demonstrate that the nexus between Simondon's ontology, his psychology, and his study of techniques could be located precisely in the notion of *mentalité technique*. Simondon himself dedicated a rich but, in his lifetime, unpublished text to this concept (Simondon 2014). My own interest in his university courses on psychology stemmed above all from the encouragement of my doctoral advisor, Prof. Fabio Minazzi, who urged me to investigate the connection between Simondon's psychological studies and his passion for technical objects. Yet, although I subsequently produced shorter essays on the subject, I never pursued a full exploration of Simondon's psychology in the manner undertaken by Sunseri in this book. I had, however, recognized its value, especially with regard to a radical rethinking of group psychotherapy for patients with psychiatric disorders, after reading several illuminating works by the psychiatrist and psychotherapist Franco Fasolo (2009; 2011). Fasolo frequently cited, in those texts, the third and fourth parts of Simondon's *thèse principale pour le Doctorat d'État ès Lettres, L'individuation à la lumière des notions de forme et d'information*, namely the volume published in France in 1989—the very month of Simondon's death—under the title *L'individuation psychique et collective*.

Sunseri's volume thus inaugurates, at least among Italian scholars of Simondon's thought, a significant and original shift of direction, one that will undoubtedly enrich the already complex secondary literature on the author with a genuinely innovative contribution. Of particular merit is Sunseri's stylistic choice in examining her subject: she offers a precise yet accessible synthesis—at once scientific and readable—of several of Simondon's writings on psychology as well as of some of the most important works critically engaging with his philosophy. Beyond her stylistic merits, certain unexpected comparisons she introduces—with authors seemingly unconnected to

Simondon's intellectual trajectory—prove to be not only insightful but also highly persuasive.

This brief *Preface*, however, is not merely intended as an invitation to read Sunseri's text—which, through both its style and its content, already compels the reader to engage with it. It also aims to draw attention to a theme that emerges within its pages and that is of pressing relevance, not only philosophically: the theme of otherness. Sunseri raises this issue—rightly so—without developing it further, for clear programmatic reasons, in relation to the conceptual and social character of technical objects. It is worth recalling that in the oft-cited text examined by Sunseri in the first of the four chapters of this volume—written by Simondon in 1956–1957 in view of its publication, in abbreviated form, in the article co-authored with Le Terrier for the *Histoire de la Science* volume of the *Encyclopédie La Pléiade* (Simondon, Le Terrier 1957, pp. 1449–1456), and taken up again in a contemporary university course at Poitiers—Simondon defined the technical object, from a social perspective, as a *socius* of human beings. It is, above all, an ally in their public activities, rather than in private life (Simondon 2015, pp. 21–249). Although technical objects may be studied in their evolutionary processes through categories borrowed from biology, they cannot, as Simondon clarified in his *thèse complémentaire, Du mode d'existence des objets techniques* (1958), be "naturalized"—that is, reduced *sic et simpliciter* to the structures and dynamics of living beings.

It is precisely from this intrinsic difference between the living and the technical—and at the same time from their discontinuous but inseparable connection—that the challenge of the otherness of the technical object in relation to the human arises. Whereas many scholars of animal life, and particularly advocates of animal rights, have occasionally addressed the question of the animal as the Other (a theme already powerfully articulated by Jacques Derrida (2025)), the otherness of technical objects has been less explored. Here, however, the risk is not one of the failure but rather one of the potentialities. In this sense, the recent "revolution" in the widespread use of artificial intelligence—even among non-specialists—has served, one might finally say, as a battering ram to force recognition of this challenge. In fact, this "new technology" has been debated since at least the 1980s, as evidenced by a colloquium held in 1987 by the *Société de philosophie française*, in which Gilbert Simondon himself participated discreetly (Gardin 1987, pp. 70–112).

If we were to simplify greatly—at the risk of some oversimplification—the alterity and the challenge at stake here lie above all in the different ways AI and humans "think", assuming the latter can even be generalized. Of course, we know AI does not literally think. What matters, however, is that its procedures employ algorithms fundamentally different from those of human cognition. Once again, this is a distinctly Simondonian question: the challenge consists in creating a dialogue—across an irreducible gap—between two heterogeneous modes of functioning, each "other" to the other.

How this relates to AI is already the subject of study. As Turing observed, AI is the realization of a "machine that learns", learning especially from its contact with human cognition. Yet, as with every genuine technical system, this virtuous circle also feeds back into the other term of the relation. As Simondon himself emphasized, the relation precedes the terms that enter into it; they are products of the relation rather than independent entities simply placed in relation—they are, in other words, nodes within a web of relations.

A broad new field of inquiry is thus opening—one might paraphrase Galileo's Discourses and Mathematical Demonstrations Relating to Two New Sciences—particularly in psychology and, more specifically, in experimental psychology, the very field in which Simondon trained and specialized. To keep pace with transformations of the human, psychology can no longer avoid confronting this decisive issue: how will the human being of tomorrow think, when human cognition learns from AI just as AI learns from it?

In this sense, the theoretical orientation adopted by Sunseri in her reinterpretation of Simondon proves especially timely. For the ultimate, ever-renewed challenge facing humanity has always been, in Kantian terms, to respond to the question: Who is the human being? As Simondon himself reminds us in his article with Le Terrier, perhaps a psychology of functions and operations—ever more open to the alterity of new technologies—will eventually echo Simondon's own lapidary formulation: *l'homme est... en se faisant*.

Giovanni Carrozzini
Centro Internazionale Insubrico Carlo
Cattaneo e Giulio Preti
Università degli Studi dell'Insubria
Varese, Italia

References

Bardin A., *Epistemology and Political Philosophy in Gilbert Simondon. Individuation, Technics, Social Systems*, Springer, New York 2015.

Barthélémy J.-H., *La philosophie du paradoxe: Prolégomènes à la Relativité philosophique*, Éditions Matériologiques, Paris 2025.

Barthélémy J.-H., *La Société de l'invention: Pour une architectonique philosophique de l'âge écologique*, Éditions Matériologiques, Paris 2018.

Barthélémy J.-H., *Manifeste pour l'écologie humaine*, Éditions Actes Sud, Paris 2022.

Bontems V., *Bachelard*, Les Belles Lettres, Paris 2010.

Carrozzini G., *Gilbert Simondon filosofo della mentalité technique*, Mimesis/Centro Internazionale Insubrico, Milano-Udine 2011.

Carrozzini G., *Gilbert Simondon: brève histoire d'une réception difficile*, https://www.implications-philosophiques.org/gilbert-simondon-breve-histoire-dune- reception-difficile/ [2018].

Carrozzini G., *Gilbert Simondon: brève histoire d'une réception difficile*, «*texto! Textes & Cultures*», XXIX, n 1–2, 2024, Coordonné par Créola Thénault Baltaretu, http://www.revuetexto.net/index.php/docannexe/file/88/Parutions/Marges/docannexe/file/104/Parutions/Marges/images/docannexe/file/2459/index.php?id=4992.

Derrida J., *L'animal que donc je suis*, édité par Marie-Louise Mallet, Folio, Paris 2025.

Fasolo F., *Gruppoanalisi e slaute mentale*, CLEUP, Padova 2009.

Fasolo F., *Respirare il gruppo. Introduzione alla gruppoanalisi senza neutralità*, CLEUP, Padova 2011.

Gardin J.-C., *Questions d'épistémologie pratique dans les perspectives de l'intelligence artificielle*, «Bulletin de la Société française de Philosophie», séance du 28 mars 1987, 81e Année, n. 3, juillet-septembre 1987, pp. 70–112.

Menegalle G., *Simondon's Heterodox Naturalism: Form, Information and Self-Organization in Imagination and Invention*, «Paragraph», XLVIII, n. 3, 2024, pp. 341–358.

Acknowledgments

I am deeply grateful to the many people who have contributed in different ways to the research presented here. These have been intense years, beginning during a pandemic that has disrupted lives around the world. I would like to express my deepest gratitude to Prof. Andrea Le Moli, my doctoral supervisor, who has believed in my research from the outset. He has been, and continues to be, an invaluable guide, teaching me the fundamental values and practices of conducting research on a daily basis, all the while being a great teacher to his students. He always says that 'nothing is thrown away' because we are all made up of mistakes and deviations that cannot be avoided or eliminated; however, we can recognise the beautiful and formidable practice of creativity within them.

We would like to express our sincere gratitude to Giovanni Carrozzini, Cécile Malaspina, Pascal Chabot, Jean-Hugues Barthélémy and Andrea Bardin for generously sharing their insights on Simondon and for collaborating with our research group on the event organised to celebrate the centenary of Simondon's birth in Palermo in 2024.

Special thanks go to Prof. Rosaria Caldarone for her constant support and esteem; to Profs. Fabio Grigenti and Gaetano Chiurazzi for their valuable comments, which enabled me to improve this work; and finally, to Prof. Marcello Di Paola for our stimulating discussions on Simondon's most complex themes and for his help with the English translation of this work.

I would like to express my sincere gratitude to the Neocybernetics Collective (*CCCC*) for welcoming me onto the board and broadening my horizons: Special thanks go to Brunella Antomarini, Marko Ćeranić, Kicca D'Ercole, Saverio Forestiero, Nicola Lettieri, Alessandro Pluchino and Daniele Poccia.

I would like to express my heartfelt gratitude to my family for their unwavering support throughout my journey, especially after I chose to study philosophy.

Thank you to Stefania for sharing every moment of this journey with me, for being both the analytical and the crazy part of me, and for being the photograph of my future. Remember that we still have so much to do together.

Thank you to Riccardo for recommending Simondon's first book to me and telling me that I would find the answers to my strange questions there. You have never stopped stimulating my curiosity, and I hope we will continue to share research on our "French friends".

Thank you to Mari for being there from the very beginning and for always believing in me. Your support made all the difference.

Finally, I would like to thank all my friends who always feel like "home", even when we are unable to stay in touch as often as we would like in these strange times. It is through this that I have come to fully understand Simondon's concept of the collective.

Contents

1 **Introduction** 1

2 **Gilbert Simondon's Genetic Psychology** 7
2.1 Introduction 7
2.2 Towards the Licence En Psychologie 8
2.3 Between Philosophy and Psychology 10
2.3.1 Merleau-Ponty 10
2.3.2 Henri Piéron 12
2.4 Between Tours and Poitiers 14
2.4.1 Gaston Bachelard 14
2.4.2 Paul Fraisse 16
2.5 1953–54: Psychology and Cybernetics 17
2.6 1956–57: The Proposal for a Genetic Psychology 22
2.6.1 First Phase: Causal Dualism 23
2.6.2 Second Phase: Systematic Monism 26
2.6.3 Third Phase: Genetic Pluralism 34
2.7 The *Change of Focus* with Georges Canguilhem 39
2.8 1958: Reconceptualising Reality Between *MEOT* and *ILFI* 42
2.8.1 Du Mode D'existence Des Objects Techniques 47
2.8.2 Psychic Individuation 51
2.9 Forme, Information, Potentiels 62
References 69

3 **The Courses of the 1960s–1970s** 73
3.1 Introduction 73
3.2 Imagination and Invention 1965–1966 74
3.2.1 Motor Anticipation 76
3.2.2 The Infra-Perceptual Image 80
3.2.3 The Image-Symbol 83
3.2.4 Invention 87
3.3 Conclusion: A Conceptual Laboratory for a Theory with *Psyche* 92
References 93

4 The Reception of Simondonian Psychology 95
4.1 Introduction: The First Debates in France 95
4.2 Bernard Stiegler and Epiphylogenesis 100
4.3 Pascal Chabot and *l'œuvre Au Noir* Di Simondon 102
4.4 Jean-Pierre Esquenazi and Affectivity as a Hinge of Individuations 104
4.5 Jean-Hugues Barthélémy and Anxiety Between Simondon and Heidegger 106
4.6 The Italian Reception: Between Paolo Virno and Giovanni Carrozzini 108
4.7 The Anglo-Saxon Reception: Tucker's *Affective Technological Consciousness* and Celis Bueno and Schettini's *Transindividual Affect* 111
4.7.1 A. Tucker's Affective Technological Consciousness 111
4.7.2 B. Celis Bueno and Schettini's Transindividual Affect 116
References 117

5 Conclusions 119

Chapter 1
Introduction

This research began with the intention of tracing in Gilbert Simondon's thought an encounter between the theories of information, cybernetics, and complex systems on the one hand, and recent evolutionary theses that are increasingly moving in the direction of possible relational ontologies on the other. In the first instance, therefore, Simondon's main works were investigated: *ILFI* and *MEOT* to get a clearer picture of the implications of one on the other. In this way, the profound connection that exists in the Simondonian approach between technique and being, as well as between technique and nature, emerged. In a 1960 text discussed at the *Société française de philosophie* Simondon, stated that he wanted to investigate the technical object, its genesis and evolution (*MEOT*) as a paradigm for tackling the theme of individuation starting from a reformed approach to the problem itself: no longer putting the individuated in the foreground but rather the process that enabled its individuation. At the same time, research into Norbert Wiener's cybernetics and the theory of complex systems was deepened in order to verify the consonances between these theories and Simondon's with regard to the relationship between the technical and the biological found in the parallel reading of his two doctoral theses. In the light of these insights, several points of contact and differences were found between the Simondon of the 1950s (the one that emerges from the theses) and Norbert Wiener's perspective proposed in the 1948 volume. On the one hand, Simondon's encyclopaedic vocation is clearly inspired by the Wienerian idea of cybernetic groups as ensembles of specialists from different fields working together to find a single thesis to explain several phenomena. On the other hand, however, the structural analogy between machines and organisms and the feedback system proposed by Wiener do not appear to be adequate solutions to fit into Simondon's relational ontological framework.

With respect to systems theory Simondon's perspective totally espouses the idea of the primacy of relationship over individuality, allowing him to overcome the various dualisms (form/matter, body/soul, nature/culture, technique/nature) that he saw deeply rooted in the French philosophical culture contemporary to him. At that

F. Sunseri, *Psychology and Technology in Gilbert Simondon*, UNIPA Springer Series,
https://doi.org/10.1007/978-3-032-25740-6_1

time, French academic culture was focused on the debate on Descartes and Hegel. As Gilles Deleuze outlines in the same years of Simondonian education, it was, therefore, necessary to change perspective and elaborate new theories (Deleuze 1953).

One of the Simondonian texts that most clarified the author's position with respect to cybernetics and complex systems theory is a contribution presented in 1962 during the conference on the concept of information organised in Royamount with the support of Maurice Merleau-Ponty. Entitled *L'amplification dans les processus d'information*, it was subsequently edited and included in the collection *Communication et information*. It is a text of particular interest because within it Simondon specified what he meant by *information* (opposing Wiener) and *transduction*, two key terms within his discourse. It was precisely the definition of transduction as a process that manages to hold together modulation and organisation that served as a link with the theory of complexity proposed in those years by, among others, Francois Jacob, Jacques Monod, Ilya Prigogine, Isabelle Stengers and Edgar Morin, in which he proposed to refound the assumptions of science by abandoning biologism and anthropocentrism; a thesis that Simondon had already placed at the basis of his research starting from the works of his master, Georges Canguilhem, in the sense of an *axiomatics of knowledge* whose method involves the study of the relationships between things, in particular through *allagmatics*, i.e. the study of conversions or operations.

My in-depth study of Simondon's theses has then shifted to the collections, texts that are less well known to the general public but contain the necessary links to detect the originality of Simondon's perspective. These are collections of transcripts of articles, courses, lectures and interviews held by the author between 1950 and the 1980s, from which it was possible to extract the evolution of his thought and the intertwining with other disciplines such as cybernetics, psychology and sociology. The result was an encyclopaedic Simondon scholar endowed with a curious and enthusiastic gaze towards the sciences and, above all, towards those disciplines whose object of study is, by its very essence, plural. The texts contained in these volumes have reinforced the awareness that Simondon's two doctoral theses are intimately connected and that it is only from an overview that the revolutionary nature of his philosophy can be traced.

The content of these texts has been of fundamental importance in tracing a network of interactions that Simondon had with scholars contemporary to him such as Raymond Ruyer, whose thought is situated precisely in the same intersection between philosophy of technique and philosophy of nature in which the author positioned himself and whose consequences, however, veered towards a teleology from which Simondon remained distant. Moreover, the influence of Georges Canguilhem's teaching regarding the possibility of finding a way to hold vitalism and mechanism together without falling into mere scientism is undoubted. Starting from Canguilhem's theses, Simondon grasped the idea of conceiving of the organism not on the basis of mechanical models that would reduce it to a machine, but rather of considering it from the angle of its relationship with the environment in which it lives in constant relation to the «errors» or «abnormalities» in this environment, and being something more than the sum of its parts.

With a view to a wide-ranging illumination of the various aspects of this encyclopaedic weaving, the analysis of the texts contained in the collections has inevitably oriented research towards an aspect that, until recently, critics had considered secondary. And which in reality, on closer analysis, turns out to be decisive in the construction of the project from the very first years of its elaboration. It is the interest developed by Simondon in psychology since his years at the École Normale Supérieure and then matured during the period of PhD thesis writing (*ILFI* and *MEOT*).

The intention of my work is thus to emphasise the interdisciplinary vocation of Simondon's reflection to demonstrate how it is no coincidence, or mere «hasard universitaire», that Simondon first became a professor of general psychology rather than philosophy. This demonstration starts from the relevance of the topics covered within the courses Simondon taught before 1958 and then in the 1960s and 1970s, as well as from the interweaving that can be traced between doctoral theses, articles and essays published in the 1950s. But above all, the relevance of psychology within the Simondonian conceptual infrastructure also emerges from the role that psychic individuation plays in *ILFI*. As will be made clear in the second chapter, psychic individuation is the point of connection between living individuation and collective individuation, and it is precisely the structuring of perception and affectivity that allows the living individual to continue his path of individuation towards a new phase of connection with the pre-individual to which he has always been bound. In the absence of perception and affectivity, the living individual is unable to look at the other as, in the first instance, different and, in the second instance, fundamentally part of the same processuality of being (the preindividual and the collective). Transindividuality develops and is fully realised in psychic individuation and enables the reconnection of all the individuated (bound by the pre-individual) in the collective.

It could also be seen that the psychological question, so deeply interconnected with philosophy, also takes root in that reflection on technique that made Simondon's thought revolutionary. In *MEOT*, the interest in the mode of existence of technical objects stems from Simondon's need to break down the system of rational control that generated the belief that humans represent the rational part of the living and, therefore, can and must keep all others under control. In this way, technical objects, mere agglomerates of matter that follow the rules dictated by their makers, have never had the right to be part of culture. In opposition to this tendency further reinforced first by the Enlightenment and then by Positivism, Simondon proposes giving technical objects an ontological place. In this direction, the Simondonian psychological «revolution» is instinctual/affective: man must have a caring relationship with technical objects, in the sense of recognising a common trait that it shares with them and resides in a primitive operativity. The nerve centre of the relationship between men and machines is therefore no longer control and automatism, but the regulation and creativity inherent in what Simondon identifies as the real secret of successful (living and technological) innovation: dis-automatism.

Simondon thus re-proposes, in his two 1958 theses, the intention of founding a genetic psychology (already conceived in in 1956–1957) in the sense of an interdisciplinary study that focuses attention on the genesis process in all dimensions of reality. For him, the different phases of the real (physical, living, psychic, social and technical) are different levels of the same process.

From 1960 onwards, Simondon's interest in psychology intensified, as evidenced by the paper he presented to the Société française de philosophie (*Forme, information, potentiels*). In the latter, psychology and sociology are precisely the disciplines that serve as examples to demonstrate the need for ontological, epistemological and gnoseological reform in the direction of a profound trans-connection between the dimensions of being. Through the courses of general psychology, a true psychology is beginning to emerge that takes into account the different aspects of a genetic study of the psychic strata. To illuminate this aspect, which the present work barely scratches the surface of, and I intend to deepen and structure in the future, emphasis has been placed on the course dedicated to imagination and invention of 1965–1966, given the undeniable connections found between imagination, the process of individuation (*ILFI*) and that of materialisation (*MEOT*). This is a first step in the direction of a re-evaluation of the entire Simondonian psychological journey, the continuation of which should consist of first and foremost in tidying up the material available up to the 1980s in order to make the importance of psychology in the conceptual reform proposed by Simondon clearer and more evident. With a view to this renewed interest in the psychological aspects of Simondon's thought and their implications for the classical areas of analysis of his work (ontology and technique), the existing critical literature should also be reviewed. The fourth and final chapter of the thesis is therefore dedicated to the network of reflections on Simondonian thought that have followed one another since 1989 (the year of Simondon's untimely death) with a focus on works devoted to the psychological question. This is a fairly small quantity compared to the number of texts focusing on the question of technique or generally on individuation. The reflections collected, however, are decisive for the thesis underlying this work. It will be clear from the texts analysed that above all the question of affect as the central junction of psychic stratification is fundamental to replacing the paradigm of the primacy of perception. Affection organises and structures the individual and his relationship with otherness, and it is thanks to it that the collective dimension is accessed. Insisting on Simondon's conception of affectivity makes it possible, among other things, to rethink the instance of control criticised in *MEOT*. Affection, in fact, is a force that attracts and repels, but it is so because it draws directly from the pool of energetic potentials: the pre-individual. Control, then, is transformed into regulation of potentials that cannot be organised and structured except when they are acted upon.

In more recent reflections, technology, information and society become the concepts with which Simondon's idea of affectivity interacts most. Extending psychological processes and affectivity beyond the individual human body, the analysis to be pursued is that of a relationship, firstly with the part of ourselves that we do not think belongs to us (the indeterminate, the irrational) and, secondly, with the other. Affectivity, therefore, becomes the force that drives the relationship from which every process of individuation (be it physical, living, psychic, social or technical) is triggered. Giving value back to affectivity as Simondon did is also at the basis of the project of the posthuman reflection of the affective turn. In it, as will be seen, several points present in the Simondonian thesis come back into play, which could open up reflections on the relationship with the new digital technologies (which Simondon never dealt with) that call attention to the affective/desiderative sphere itself.

Chapter 2
Gilbert Simondon's Genetic Psychology

2.1 Introduction

This first chapter provides a reconstruction of the genesis and evolution of Gilbert Simondon's proposal for a *psychologie génétique*, a pivotal yet under-explored facet of his philosophical oeuvre. Psychology is not a marginal concern in Simondon's thought; rather, it plays a fundamental role. Indeed, it serves as a transversal axis through which it is possible to understand the individuation of the physical, biological, technical and psychic in their mutual resonance.

The subsequent discussion will retrace Simondon's early academic career—from his formative years in psychology, to his dialogue with key figures such as Merleau-Ponty, Piéron, Bachelard and Fraisse, to the decisive influence of Canguilhem. The chapter will then outline the progressive consolidation of a genetic approach to psychic life. This evolution culminates in a series of key moments, including his encounter with cybernetics in the early 1950s, the formulation of a three-phase model of psychological theorisation between 1956 and 1957 (causal dualism, systematic monism, genetic pluralism) and the parallel maturation of his ontological framework between *Du mode d'existence des objets techniques* ([22] *MEOT*) and *L'individuation à la lumière des notions de forme et d'information* ([89] *ILFI*).

The chapter demonstrates that, rather than isolating psychology from Simondon's broader project, his genetic psychology in fact anticipates and supports the very possibility of a general theory of individuation. A thoroughgoing analysis of unpublished lectures, manuscripts and published texts reveals that Simondon's innovation lies in his ability to articulate psychological processes, in particular affectivity, imagination and invention, as key vectors in the genesis of subjectivity and collective becoming. The chapter culminates in an exhaustive examination of *Forme, information, potentiels*, wherein Simondon delineates the outlines of an allagmatic approach that seamlessly integrates psychic individuation into a comprehensive philosophical and transdisciplinary framework.

F. Sunseri, *Psychology and Technology in Gilbert Simondon*, UNIPA Springer Series,
https://doi.org/10.1007/978-3-032-25740-6_2

2.2 Towards the Licence En Psychologie

In 1946, a young Gilbert Simondon obtained the *Licence en Philosophie*, the *certificat d'Études Supérieures en Psychologie* and that in *Histoire de la Philosophie* [1, p. 20]. During his preparatory years, he had the opportunity to meet for the first time the president of the UIPS (*Union International de psychologie scientifique*) Henri Piéron, whose course in the psychophysiology of sensations he had taken at the Faculty of Science of the ENS (*École Normale Supérieure*) in Paris. In the mid-1940s, Simondon had been able to move to Paris after attending the ENS preparatory courses held in Lyon due to the war. He had therefore had the chance to attend the courses of Martial Gueroult (who would later follow him for his thesis for his higher diploma in 1948), Maurice Merleau-Ponty, Jean Hyppolite, Jean-Toussaint De Santi, Georges Gusdorf, Jean Laporte and Jean Wahl. While pursuing his first vocation, philosophy, which had led him to clash with his father who wanted him to be an engineer instead, he had become increasingly interested in the scientific and psychological worlds. In particular, in 1947 he had obtained a *certificate* in mineralogy from the Faculty of Science in Paris and a *certificate* in psychophysiology under Alfred Fessard, one of the founders of neuroscience in France.

It was precisely these studies in psychophysiology that led Simondon, at the suggestion of Gaston Bachelard, to write four letters to Henri Piéron [2, pp. 101–103] in 1950 in which he asked to continue his studies in the psychophysiology of perception at the Sorbonne laboratories. Simondon's interest in psychology was strengthened during his years of teaching at the Lycée Descartes in Tours, a post he obtained in 1948 following his *agrégation* in philosophy. In addition to teaching philosophy, he established a technology laboratory at the school where students could interact with technical objects and understand how they functioned. It was within the confines of this laboratory that Simondon initiated the formulation of the theory positing a robust correlation between technical and psychological research. His stance on the approach to psychology is revealed in his correspondence with Piéron, where he asserts that the study of behaviour must take into account the transformative procedures that individuals carry out in context. The discourse with Piéron was accompanied by courses in psychophysiology, child psychology and social psychology, and it was this exchange that led Simondon to obtain the *licence de psychologie*, which had been instituted in France a few years earlier at the insistence of Piéron, who wished for psychology to be recognised as an autonomous science with respect to philosophy [1, p. 24].

Two years later, in 1952, Simondon made the decision to pursue academic studies in the United States, at the University of Minneapolis. Here, he enrolled in a three-month course in social psychology. Following his return to France, he attended the seminar on experimental psychology held by Paul Fraisse [3, p. 15]. During this seminar, Simondon presented a paper on different psychological models. Fraisse had become a key figure in the French academy because, from his position as an experimental psychologist, he was attempting a profound transformation of the perspective on psychology while clashing with clinicians, such as Piéron, who frowned upon this new approach. It was Fraisse himself, several years later (in 1963), who would support the candidature and victory (against Didier Anzieu, the clinicians' candidate) of Gilbert Simondon as professor for the nascent chair of general psychology at Paris V: a profound rupture between the experimentalists and the clinicians that would lead, in 1968–1970, to a real split [4].

After Fraisse's course, Simondon would publicly disclose his position in two articles published in *Cahiers pédagogiques* in November 1953 [5, pp. 165–190] and then in October 1954 [5, pp. 191–208], in which he would report on the experience of technical initiation carried out with the fifth class of the Lycée Descartes. The report demonstrates Simondon's perspective on the interconnected nature of psychology, philosophy and technology, emphasising the importance of comprehending structures (vital and technical) from the perspective of their genesis. In this workshop, Simondon eschewed theoretical lectures, apart from a few preliminary outlines. Instead, he divided the students into groups, who were tasked with constructing the various components of a technical object, subsequently assembling it, and finally testing its functionality. The traditional organisational framework of secondary education, which historically equipped students for roles in fields such as administration or engineering, underwent a paradigm shift. Instead of focusing on preparing individuals for specific professions, it introduced a new focus on the theoretical underpinnings of technological entities and the processes that shape them. According to Simondon, this shift facilitated the establishment of a profound and transformative connection between students and their learning material.

This marked the commencement of a pedagogical approach to psychology by Simondon that many of his colleagues, including Fraisse himself, would never fully comprehend. The pedagogy he employed was interdisciplinary, encompassing philosophy, science and technique, and it was characterised by an unwillingness to be confined to a particular school of thought. This approach was already evident in the classrooms of the Faculty of Humanities at Poitiers between 1950 and 1963. During this period, Simondon experienced several clashes with his peers due to his interdisciplinary approach and his enthusiasm for technical experimentation, which transformed the faculty's corridors into a kind of open-air workshop. In these spaces, Simondon sought potential technological innovations for his Peugeot [1, 6]. The period was characterised by a high level of academic productivity, with Simondon establishing a psychology laboratory and a technology laboratory, in addition to dedicating himself to the structuring of the two doctoral theses he presented in 1958.

2.3 Between Philosophy and Psychology

2.3.1 *Merleau-Ponty*

Carrozzini characterises Simondon as a «Socrates of our times» [7, p. 7],[1] a figure prone to anger, especially in the face of opposition from those who, due to their inability to comprehend his encyclopaedic and articulate vision of knowledge, seek to contradict him. During his interview with Jean Le Moyne in 1968, Simondon is depicted as donning a mackintosh over a smart suit [7, p. 7], seemingly downplaying the extent of his thought. He entrusts the interweaving of philosophy and technology to chance, a decision that could be seen as a deliberate strategy to avoid oversimplification. Simondon's reluctance to engage in ostentatious displays or engage in social interactions is attributed, at least in part, to the existential angst that those who knew him perceived in his character. This tendency was also possibly influenced by his rural upbringing, which likely contributed to his disconnection from the social milieu of metropolitan areas. Moreover, the Rhône-Alpes region, where Saint-Étienne, Simondon's hometown, is located, at the time represented a surreal scenario, between industrialisation and rituality. It was in this context that the philosopher began to develop a thought that could link precisely these two aspects: technique and magic [7, pp. 8–9].

Simondon's academic background was primarily in the humanities, with several notable distinctions for his philosophical dissertations [7, pp. 10–11]. However, he also demonstrated a high level of aptitude in scientific disciplines such as physics and chemistry. His interest in technology, and more specifically, mechanics and electronics, was not piqued within the school environment, but rather through field observations in his rural locale. His father, recognizing his aptitude for technical objects, had anticipated a career for him as an engineer,however, his interests were directed towards different pursuits.

His entry into the École Nationale Supérieure (ENS) in Paris in 1945, following a preparatory course in Lyon, marked a pivotal moment in the formation and structuring of his thinking. From '45 to'48, he had the privilege of studying under renowned figures such as Maurice Merleau-Ponty and Henri Piéron, professor of the psychophysiology of sensations at the Faculty of Science.

Between 1947 and 1948, the years during which Simondon sat on the benches of the ENS, Merleau-Ponty gave a course on the union of soul and body in Malebranche, Biran and Bergson, the transcripts of which would not be made public until 1968 [8]. In this course Merleau-Ponty reads the authors through the lens of that philosophy of the body that accompanied his philosophical production until his tragic death in 1961. In Merleau-Ponty's philosophy, starting as early as *The Structure of Behaviour* (1942) and *The Phenomenology of Perception* (1945), corporeality is fundamental

[1] Translation mine. Henceforth, all English translations of works, for which no translation exists, are mine.

to perceiving by accessing the *Lebenswelt*, the concrete world of life of Husserlian derivation. In opposition to Husserl's position on *noesi-noema* (act of thinking and intentional object of thought), Merleau-Ponty proposes a more body-centred phenomenology by outlining certain phenomena in which *noesi* cannot be distinguished from *noema* such as, for example, bodily self-perception or the perception of the other. In these cases, the thinking subject and the thought object cannot really be separated or associated with a single element. Thus, the separation between an inner part, a consciousness or soul, and an outer part, a body, cannot be experienced, thus objecting to the Cartesian thesis of the separation between the two substances as apparently unavailable in the world of life. The body is, therefore, at the centre of Merleau-Pontian reflection for its being both actor and spectator of perception. A body always embodied as a mode of being and not a possession, and as or an objectif body analysable by science. These two modes of the body coexist as subject and object. In this sense, the example of the hand touching the other hand is emblematic [9–11] and, in doing so, both discover that they are always ontologically split, never alone. Being in the world of life sees itself as part of it insofar as it is made of the same texture, the same element or, in Merleau-Pontian terms, the same *chair* (flesh). This is the logic of the double *étonnement* that, according to Merleau-Ponty, is inherent to the philosophical attitude: first we are astonished by the world outside us and then by the astonishment itself when we recognise that we are part of that outside us. This faculty of always being double of the body is the great novelty brought by Merleau-Ponty to what can be called a phenomenology of the flesh, whose theoretical basis is Husserlian but whose implications open up perspectives that cannot remain within the concept of intersubjectivity. The body made of flesh, the non-objective body of science, is never alone with itself since it is always exposed to the other: in Bruno Moroncini's words, the flesh of the body « is the otherness that affects the living body from the beginning and prevents it from being only my body» [12, p. 282].

As Moroncini asserts, Merleau-Ponty is associated with the group of philosophers who, in the early decades of the twentieth century, sought to challenge the empiricist positivist theory that tried to prevail in all domains of knowledge, including psychology [12, p. 283]. In particular, in continuity with the theory of *Gestaltpsychologie* and some Bergsonian theses, the French philosopher attempted to convey to his students the importance of the relationship between the perceiver and the perceived as an object of study as much as the individual observable empirical realities. The relevance of the relationship stems from the observation of the impossibility of defining the psyche as something that is constituted prior to the encounter with the other. This is because consciousness is always conscious of the world, as opposed to being conscious of something as Husserl claimed. This means that the construction of consciousness and the attribution of consciousness necessarily passes through the body, the visible. However, the corporeal «other» is only distinguishable from the corporeal «me» at a subsequent stage, as, according to Merleau-Ponty, there is initially an «anonymous collectivity» [9, p. 91] and the development of consciousness of self and other only occurs later. This *anonymous collectivity* appears to be evocative of the concept of *collectivity* as theorised by Simondon. Nevertheless, it can be argued that Merleau-Ponty's notion of anonymous collectivity is more closely associated with the notion of trans-individuality, which, in a sense, occurs prior to the formation of collectivity. In the context of trans-individuality, the concept of the 'I' becomes indistinct, giving way to a more collective 'we', thereby facilitating the transition to collectivity. However, a fundamental and discernible distinction emerges: anonymous collectivity appears to originate from two pre-existing polarities, while trans-individuality signifies the emergence of a pre-individual that comes into consciousness through orientation in the environment, preceding any notion of an 'I'.

The development of consciousness of self and other, as posited by the philosopher, occurs through the technique of the mirror, a concept derived from Lacan. The subject, by perceiving the other as a body, comes to recognise that the other is, in turn, a seen body. In essence, Merleau-Ponty's argument suggests that the other exists prior to the self, as it enables the self to recognise itself, develop its psyche, and, concomitantly, that it is through this recognition of self as a body that the subject gains awareness of its place in the world.

Accordingly, the body occupies a pivotal position in Merleau-Ponty's philosophical discourse, particularly with regard to his emphasis on perception. In his inaugural address at the Collège de France in January 1953, he proposed that perception and sensation are essentially synonymous [10]. It is from this interest that the French philosopher dedicates a significant portion of his studies to perception, commencing with observations and clinical cases previously analysed by *Gestaltpsychologie*. This approach is further influenced by *Gestaltpsychologie* psychology theorists, as evidenced by Merleau-Ponty's notable engagement with the study of perception as an essential component in the structuring of consciousness.

The profound impact of these theses on a young Simondon is also evident, as evidenced by his discussion of his doctoral thesis (*ILFI*) [89] in which he argued for the eradication of dualisms in order to establish a genetic ontology. These dualisms were the same ones against which Merleau-Ponty had raged.

2.3.2 Henri Piéron

Simondon's interest in psychology also emerges from his decision to take Henri Piéron's course, which enabled him to obtain the *certificat* in psychology. This decision allowed him to immerse himself in the French academic psychology environment, acquiring an understanding of its dynamics. This ultimately led to his appointment as director of the psychology laboratory, initially at Poitiers and subsequently at the Sorbonne.

Since the 1920s, Henri Piéron has been regarded as a proponent of physiological psychology in France [13, pp. 305–309]. He campaigned for the establishment of psychology institutes at universities, where the discipline was regarded as a mere appendage of philosophy. In January 1921, Piéron had succeeded in obtaining the first institute of psychology, whose statutes stipulated that psychology as a laboratory science would be studied within it, as well as the various psychological sciences that departed from the applied aspect [13, p. 305]. In a manner analogous to Watson's influence in the United States, the youthful Piéron was entrusted with the responsibility of orienting psychology towards positivism by establishing it as an independent scientific discipline [14, p. 70]. During his inaugural lecture at the EPHE at the Sorbonne in 1907, Piéron articulated his objective of exploring consciousness as a means to comprehend the psyche of the living organism, emphasising the value of sensory analysis [15, p. 291]. To this end, in addition to obtaining the institute of psychology, Piéron had moved to found and direct the journal *L'Année psychologique*, as well as the chair of physiological psychology at the Collège that would lead to his meeting with Simondon. Among the other participants in the establishment of the newly founded institute of psychology, it is crucial to acknowledge the contribution of Étienne Rabaud [14, p. 71], who held the position of lecturer in animal and comparative psychology. Rabaud would become a constant reference point for Simondon, particularly in the section of the *ILFI* dedicated to the individuation of living beings. In this regard, it is noteworthy that Simondon's interest in psychology was by no means marginal, and his profound understanding of the internal dynamics of the discipline's evolution in France significantly influenced his theoretical framework.

Notwithstanding the fact that the inauguration of the Institute of Psychology in 1921 had resulted in an increased level of importance being assigned to the discipline, Piéron advanced the argument that the resources allocated for the training of French psychologists remained negligible. The paucity of resources allocated to psychology was further emphasised by the fact that it was a discipline exclusively available to philosophy undergraduates. Conversely, Piéron's proposal emphasised the integration of psychology as a transversal discipline across various faculties, including scientific domains.

In a letter to the rector of the University of Paris in 1923 and in an article in 1926, Piéron emphasised the crucial importance of including psychology in the training of teachers and educators [16, pp. 6–8]. In this regard, Simondon's emphasis on pedagogy and its interconnection with psychology can be regarded as a testament to the initial intentions of the young Piéron.

During the interwar period, Piéron was embroiled in a contentious debate with Marcel Mauss, a renowned anthropologist who served as president of the Société de psychologie. Mauss advocated for a form of psychology that sought to study «l'homme complet» [14, p. 84], or the complete human being, inhabiting a specific space and time within the context of society. He was supported in this endeavour by Ignace Meyerson, a young medical and natural history scholar who had joined Piéron's institute. Despite lacking formal training in psychology, Meyerson demonstrated a keen interest in acquiring knowledge through the publications of *L'Année*. The fundamental distinction between Piéron and Meyerson, aside from the former's notably austere lifestyle and the latter's decidedly libertine one, pertained to the philosophical orientation they aspired to instil in the discipline of psychology in France. While Piéron was an advocate of positivism, Meyerson adopted an encyclopaedic approach to the study of psychology, a perspective influenced by his colleague Jean Piaget, who would later become known for his significant contributions to the field [14, pp. 88–90].

Following the war, Meyerson's departure from the academic scene was precipitated by ostracism, while Piéron, in collaboration with Henri Wallon and Paul Langevin, initiated the formulation of a comprehensive reform plan for the French educational system [14, pp. 99–100]. This plan, which was never implemented, envisaged compulsory schooling up to the age of eighteen and the inclusion of more manual skills within the training.

Concurrently, in recognition of the necessity to direct the efforts of those who sought a more systematic study of sociology alongside psychology, Piéron had resolved in 1944 to establish a chair of collective psychology.

The years between 1947 and 1948, during which Simondon attended Piéron's courses, were pivotal in establishing psychology as a distinct discipline in France. The implementation of this initiative, which was spearheaded by Piéron, Wallon and Lagache, was made possible by the commitment of Daniel Lagache, a psychoanalyst who advocated for a scientific psychology that was also a concrete psychology. Consequently, the French academy approved the incorporation of the new *licence* in psychology not only in Paris, but also throughout the country [14, p. 107]. Piéron's tenure as president of the Institute and the chair in physiological psychology continued until 1952, when he was succeeded by Paul Fraisse.

This concise overview of Piéron's career highlights his overtly positivist inclinations concerning the direction he aspired to impart to psychology. Consequently, it is evident that Piéron's legacy, particularly in the context of Simondon's work, was to be characterised by a discernible critical foundation. However, Piéron's emphasis on physiology and biology as fundamental disciplines for psychology proved instrumental in fostering connections between figures such as Rabaud, Bachelard, Mayerson, and Fraisse with the French academy, whose more open-ended approaches and wide-ranging interests would subsequently facilitate Simondon's appointment as chair of general psychology.

2.4 Between Tours and Poitiers

2.4.1 Gaston Bachelard

In 1948, following his attainment of the *agrégation* in philosophy with Gilles Deleuze, Simondon was dispatched to the city of Tours in order to assume a teaching position at the Lycée Descartes. A few months after commencing his assignment, he submitted an application to be selected as an assistant to Gaston Bacherlard at the Sorbonne. Simondon's interest in psychology persisted during this period, and he successfully obtained a *licence* in the subject in 1951, three years after his institution. This was facilitated by the compulsory additional examinations in child psychology and social psychology. The following year, he attended a training course in social psychology in Minneapolis for a period of three months, after which he returned to Paris to attend the experimental psychology seminar of the young Paul Fraisse, who had succeeded Piéron. Following the experimental years spent in Tours, in 1955 he obtained the position of assistant professor of psychology at the University of Poitiers and the concomitant permission to install, at the institute of psychology, a laboratory of experimental psychology with which he had come into contact during Fraisse's seminar.

While the extent to which Simondon and Bachelard exchanged ideas is not fully documented, it is widely accepted that the Saint-Étienne philosopher's understanding of epistemology was significantly influenced by the seminal contributions of the father of French epistemology. As Jean Hyppolite, Simondon's doctoral thesis director, stated in a 1964 article in homage to Bachelard, «there was nothing decisive, nothing radical» between criticism, positivism and the epistemology proposed by Bachelard [17, p. 1]. He was, therefore, a figure who, according to the director of the ENS in Paris, had acted as a watershed within the history of philosophy by establishing a true French epistemology.

According to Bachelard, philosophy had to take note of the revolutions that had occurred in science, such as the discovery of chemistry or microphysics, to overcome the use of the concept of substance and determinism [17, p. 2]. He had announced a « a new age for Humanity» [17, p. 2] that was opening up and that required a revision of philosophical questions starting with epistemology. It was in this direction that the young Simondon seemed to be moving, at the very beginning of the 1950s, when while still in Tours and applying to become Bachelard's assistant, he proposed a «nouvel humanisme» that would include the study of technique within culture.

Hyppolite describes Bachelard as a thinker who, although capable of re-founding French epistemology by insisting on the need to separate the sciences from «naïve» sensible experience, had a great concern for the natural elements and the experience of nature [17, p. 2]. Referring back to Bergson, he would resound that *élan* and creative tendency in his texts on the psychoanalysis of the elements [18–22][2] and be able to emphasise the rupture of order and continuity that he accuses Bergson of. Creative genius is thus not to be found in the order of nature, but in the disorder present both in nature and in sciences such as arithmetic. A disorder to which Bachelard continually leads his readers, as in the case of the 1940 text *La philosophie du non* [23], in which he does not seek to affirm a philosophy of the negative, but a philosophy of variation, of the modification of the established order [24, p. 44]. Georges Canguilhem himself, former pupil of Bachelard and mentor of Gilbert Simondon, recalls the importance of this way of understanding the philosophy of non: the «philosophy of correct knowledge, philosophy of the foundation by recurrence, dialectics for Bachelard designates as a fact of culture the vector of scientific approximation whose meaning it reinforces by proposing it as a rule» [25, p. 207]. Thus, in Bachelard there is no

---

[2] It would seem that Simondon took up this Bachelardian attitude when he linked the birth and development of technology and science to the magical network that united man to nature with its elements.

Hegelian *Aufhebung*, but a process of invention of the true that is proper to science. And if the scientific process is a constantly evolving inventive making, then the reason of positivism also changes shape in Bachelard's thought and becomes what comes after science as invention and application of concepts. What Bachelard sought to bring about was an awakening of the consciences of philosophers who had slumbered under criticism, positivism and idealism, starting from the possibility of thinking of science as an inexhaustible source of philosophical questions, i.e. of concepts to be given form [24, p. 45]. In this *philosophy of non*, and in this inventive science, one cannot fail to see the beginnings of the reflection that Simondon will carry forward in *ILFI* and *MEOT*, in an attempt to propose a perspective in which technique (which is none other than Bachelardian inventive science) comes before the construction of concepts, just as in the ontogenetic process relationship precedes polarities.

It is precisely the theme of polarity that seems to have linked Bachelard to the young Simondon [26]. At the end of his time at the ENS, Simondon would seem to have started a thesis on polarity in psychology under Bachelard's direction, as his request to be assigned as his assistant would show. There is no trace of this thesis, however, as Bachelard may have misplaced the manuscript. While this last episode may have led to the estrangement between the two, it could not have eliminated the Bachelardian derivations present in Simondonian psychology.[3]

2.4.2 Paul Fraisse

On the other hand, the first contacts between Simondon and Paul Fraisse seem to date back to 1952, to the experimental psychology course. Fraisse had just taken over from Piéron, who in the meantime had retired, leaving him the task of making psychology definitively an autonomous science. Fraisse stated in his biography that he strongly believed in the unity of psychology [27, pp. 79–96] and succeeded in 1965 in having it transferred from humanities departments to natural sciences departments [14, p. 112]. Ever since his entry into the world of psychology at the end of the Second World War, Fraisse had been Piéron's new protégé, not only because he shared the same convictions, but also because of his drive for action. Personalism was emerging in those years as a challenge to reductionism, to any mechanistic conception and to any dualism. Personalists argued that materialist theses would allow psychology to free itself from the aura of spiritualism dictated by its proximity to philosophy. According to personalism, man is in fact a multiple unity that cannot be studied by fragmenting it, but only through a holistic approach [14, p. 191]. In the field of psychology, this implied studying the personality in its complexity «as a "subsisting relationship" between the facets of individual existence» [14, p. 191]. This psychology, referred to by Parot as «psychologie en première personne», was interested in the relationships between conscious phenomena and in intercepting that causal power that links two phenomena. The latter was thought of by personalist

[3] Bachelard's texts are listed in the bibliography of *Imagination and Invention*.

psychologists as the unconscious, but without acknowledging the existence of an unconscious that would work together with the conscious. The bridges between the different conscious phenomena were condensed, according to the personalists, into the personality.

Interest in personalist psychology, which originated in Germany and later became established in the United States, came to Fraisse through his great friend Emmanuel Mounier [14, pp. 200–201]. However, Fraisse's own position on personalism is not very clear, except in some statements made in the 1960s and 1980s in which he also maintains that Mounier always kept «assis entre deux chaises». In truth, Fraisse himself also seems to remain undecisive between personalism and experimental psychology, although his main interest remains the structuring of time in relation to memory [28, 29].

The figure of Fraisse does not enter Simondon's life by chance since, as Parot reports, there was a bond of friendship between Mounier and Georges Canguilhem (later to become Simondon's thesis supervisor) that had led the two to share research on the philosophy of biology [14, p. 239]. Canguilhem was very critical of biology and psychology that that would abstract from reality and consider the environment as a system of unsupported relations [30, p. 172], recalling Descartes' mechanistic assumptions. Canguilhem's position seems to espouse one of the major concerns of personalism: not being able to tackle the study of the living, in its relationship with the environment, without considering the personal relationship established between the individual and the environment with which he comes into contact. In this sense Fraisse is a complex character to pin down, torn between personalism and scientific psychology. His project, inherited from Piéron, to make psychology in France a natural science on a par with physics, biology and chemistry is poorly connected to his sympathies for his friend Mounier's theses, which proposed instead a psychology in dialogue with the person, far from laboratories. It may have been this presence of «two men» [14, pp. 201–202] in Fraisse that led to Simondon's victory in 1963, against a candidate of the clinicians, in the competition for the chair of general psychology at the Sorbonne. In the following years various professional tensions would arise between Fraisse and Simondon, as Carrozzini recounts [7, pp. 26–27], due to their different perspectives on psychology. Fraisse will always remain a follower of Piéron and the exclusivity of psychology, while Simondon, also thanks to his closeness to Canguilhem, will develop an interdisciplinary perspective ill-suited to the traditions of experimental empirical psychology.

2.5 1953–54: Psychology and Cybernetics

The years leading up to the defence of his two doctoral theses were dictated by the research and pedagogical struggle that Simondon sought to pursue within the walls of the laboratory at the Lycée Descartes, but also through articles and lectures. It is in these writings that it is possible to trace the birth of certain theories that he would then expand upon in his theses and subsequent courses.

Particularly interesting is a response to the objections raised against him by Georges Zadiou-Naïsky, professor of physics, following the article published in 1953 in the journal *Cahiers pédagogiques* entitled *Place d'une initiation technique dans une formation humaîne complète.* Simondon's answers were published in two issues, one in March 1954 and the other in October 1954. In the first issue he analysed three objections, two of which had led him to expound on the relationship between science, technology and psychology. The first objection concerned the resources he had enjoyed during his experimental teaching period at the Lycée Descartes and emphasised how it was not possible to bring about a reform in the French school system on his own strength. Simondon emphasised how strongly he believed in the need for such a reform and how his experiment at the Lycée Descartes showed how possible it was to realise "a material university", i.e. a place where structures and materials were not relegated to just a few science departments, but were common, thus opening up the possibility of school reform [5, p. 225].

Already in this first response it is possible to trace the profoundly encyclopaedic vocation that would guide all of Simondon's work. It is with the second response to the objection regarding the validity of Haeckel's law (1866)[4] that this position would be further structured. The moment Simondon cited the biogenetic law in the 1953 article, he in fact raised the ire of Zadiou-Naïsky, who considered it incorrect to use the experiment at the Lycée Descartes as verification of Haeckel's theory. Simondon would then set out a response that, far from defending Haeckel's thesis per se, rather showed the need to use this biological law in the psychological and sociological fields as a «big hypothesis» [5, p. 227] from which to draw experimental theses to work on. What he criticised Zadiou-Naïsky for was that he condemned Haeckel's law without presenting the reasons why it would be contradictory. In this brief defence, Simondon also alluded to a theory that he would later pursue during 1956 and take up in his theses: the idea that a genetic psychology was emerging in those years that was capable of bringing together psychology and sociology by comparing «the mental development of the individual [and] the individual to the genesis of society» [5, p. 227].

The final response is of paramount significance, as it pertains to the relationship between technique and science that Zadiou-Naïsky had formally endorsed as an identity. Simondon emphasised how the identity of science and technique advanced by Zadiou-Naïsky did not take into account the decisive historical passage of the writing and publication of the *Encyclopédie.* Simondon emphasised that this shift could not be disregarded, as it marked the first instance in which technique was liberated from its once enigmatic aura, rendering it accessible to the general public through the medium of the encyclopaedia. The *Encyclopédie*, as Simondon argued, presented a «system of human activity» [5, p. 229] with the capacity to revolutionise social

[4] Haeckel's law known as the Law of Biogenesis holds that ontogeny, i.e. the individual development of embryos, is an abbreviated and incomplete recapitulation of phylogeny, i.e. the evolutionary development of the species. Also instrumentalised by Nazism, Haeckel's law, however, has roots in Darwinism, Lamarckism and the theses proposed by Goethe and influenced various twentieth century thinkers and scientists. Haeckel's theories have recently been re-analysed to highlight their innovative aspects. Among others see [31, 32], pp. 258–276].

equilibrium. This was because, in contrast to the approach to knowledge structured by science, the *Encyclopédie* enabled the reader to discover how man operates in his daily activities.

Simondon had already dedicated a text to encyclopaedism around 1950, written for his lycée Descartes colleague Pierre Verdier, professor of literature. This text remained unpublished until 2016, the year of the publication of the collected book *Sur la philosophie*, in which it was included under the title *Les encyclopédies et l'esprit encyclopédique* [33]. In this text, Simondon identified the *encyclopaedia* as the first true schematisation of operations, thanks to a principle of organisation dictated, in turn, by an operation. This marked a significant departure from the magical world-view that had previously dominated, leading to the perception of structures as the products of an occult operational processes, that required belief without empirical evidence. The Enlightenment, with its concomitant spirit of scientific inquiry and the creation of the first encyclopaedia, represented a significant opportunity for humanity to overcome its primordial fear of the unknown, specifically the unseen forces that connected structures. This transition necessitated a shift in the medium through which knowledge was transmitted, with figurative language becoming the primary mode of expression within encyclopaedias, accompanied by visual aids such as drawings and graphs, collectively termed «schemes».

The aim of the *encyclopedia* was, in fact, to share technical knowledge that had hitherto remained esoteric, with the widest possible audience. Thanks to its use of figurative language, nothing could remain concealed or obscured behind words. Through this openness, the encyclopaedic spirit revealed the profound connection between the human and operational activity, even more so than with substance itself. The individual who wrote and read the *encyclopedia* embodied an interdisciplinary and collaborative spirit, oriented toward the operation that binds structures together rather than toward the structures themselves. Indeed, the *encyclopedia* did not define things of structures; rather, it described processes through the use of schemes.

It was through the encyclopaedic spirit inaugurated by the Enlightenment that it became possible to overcome the inertia imposed on society by the extreme automatism induced by occultism. Such a transformation could occur whenever a society reached its limit of automation—a theme that Simondon would later revisit in *MEOT*.

In his writings from 1954, Simondon initiated the exploration of a theme that would become pivotal to his theoretical framework: the distinction between science and technology. Through an analysis of this difference, he entered a highly charged debate in France at the time—the debate on cybernetics.[5] To fully comprehend the significance of the claims Simondon puts forward in the responses appended to his 1954 article, it is essential to compare them with a text written in 1953 and published posthumously in 2016: *Épistémologie de la cybernétique* [33, p. 177–199].

In his 1954 responses, Simondon advances the argument that «technology constitutes [...] a synthesis of a large number of sciences», and that «the technician [...] thinks teleologically, performs dynamic syntheses» [33, pp. 178–179], in contrast to the scientist and the scientific, who focus on a specific object within a single domain.

[5] Simondon's first text on cybernetics is actually dated 1953 [33].

This perspective is further elaborated in the 1953 text, where Simondon engages with the question of whether cybernetics, which had emerged five years prior thanks to the contributions of Norbert Wiener, could be regarded as a scientific discipline. In this text, Simondon asserts that science, as defined by positivism, which—with its claim to objectivity—dominated the academic sphere at the time, is not concerned with the study of operations, but rather with the study of structures, to which it applies technical operations that it does not itself examine. In effect, science directs its gaze toward a specific object—the phenomenon—upon which the subject applies either operative or reflective capacity [33, pp. 178–179].

Now, if science is considered solely from a positivist perspective, then cybernetics cannot be regarded as a science, since—as Wiener had already emphasized in 1948—it does not concern itself with the study of a structure, but rather with the operations that enable the transformation of one structure into another. In his 1954 responses, Simondon affirms that «cybernetics assumes the role of an instrument of synthesis» [33, p. 183] among the various objective sciences, a common methodological ground that would allow for the understanding of the operative connections that exist between the sciences.

Accordingly, general cybernetics—or, as Simondon had already begun to refer to it in the 1953 text, *general allagmatics*—would be tasked with providing a set of operations and definitions, along with «a certain conceptual instrumentation» that could be *transduced* from one field of knowledge to another. A *transducer* is a device that has the ability to convert one physical quantity into another by changing some of the characteristics that identify it. The notion of *transduction* had already been used by Piaget in the sense of a mental operation on a par with induction and deduction [34, p. 185]. Simondon takes it up but gives it a very different twist. Simondonian transduction designates the process of individuation itself or «an operation, physical, biological, mental and social, by means of which an activitỳ is progressively propagated within a certain domain» (*ILFI*, p. 31). It is, therefore, an operation of amplification of a condition from one level of reality to another. It is precisely the concept of transduction that allows Simondon to rethink the analogy no longer between structures, but between operations. The example Simondon gives to understand what a transduction is is the process of crystallisation. For more details, see the passages in his doctoral thesis (*ILFI*, pp. 31–35).

Rather than conceiving of cybernetics as a science, Simondon proposes viewing it as an inter-scientific technique, one that elevates technology to the level of «intellectually valid» knowledge, in contrast to the merely utilitarian sphere to which it had been traditionally confined.

This cybernetic logic—still to be fully structured [33, pp. 198–199]—was seen as capable of raising disciplines such as psychology and sociology to the status of scientifically legitimate forms of knowledge, despite the fact that their objects of study were not unified. Simondon emphasized that neither of these two disciplines had, up to that point, succeeded—despite considerable efforts—in defining themselves as sciences in the positivist sense of objectivity. This is because in neither sociology nor psychology is it possible to direct inquiry toward the isolated objects of their study—society and psyche—but only toward the relationships that these objects maintain with other realities (e.g., psychophysiology or sociogeography). Simondon hoped that, supported by cybernetic theory, psychology and sociology could legitimize themselves as sciences of more complex operations—namely, the psychic and the social—thus definitively abandoning the attempt to conform to positivist objectivism. An extensive articulation of this Simondonian position with regard to psychology can be found in his work from 1956 and in the 1960 essay *Forme, information et potentiels* (which will be analyzed in the following sections).

Also in 1954, Simondon wrote a second article, entitled *Réflexions préalables à une refonte de l'enseignement*, in response to critiques from Zadiou-Naïsky and a group of intellectuals who, at the time, advocated for a reform of the French educational system in the direction of increasing specialization. Like the 1953 article, this piece was published in *Cahiers pédagogiques* and later included in the collected volume *Sur la technique* [5, pp. 233–254].

It is crucial to recognize within this text the deep connection Simondon was drawing between the pedagogical problem France was confronting at the time and the broader issue of the need for a shift in perspective regarding technology and its object. Here, the problem of the segregation of units also begins to take shape, a theme he would later take up in his principal doctoral thesis. From the opening lines, Simondon states that the effort to be made is that of freeing educational institutions from a rigid division into levels and specializations [5, p. 233].

The problem of extreme specialization and division had been established historically with the transition in the nineteenth century from an agricultural society to an industrial one. The latter required individuals specialized in a specific kind of work, to be performed continuously throughout the year within the industrial system, whereas in the agricultural world, workers needed the broadest possible range of knowledge in order to perform different tasks depending on the needs of each season. This transformation had represented a major social shift, but it had not caused significant problems. The complications began to emerge with the advent of thermodynamics and the development of new forms of energy (such as electricity), which could be distributed across multiple structures while maintaining the same level of efficiency.

With the emergence of new energy forms and new tools for information exchange (radio, television, etc.), the vertical structure that specialization had imposed upon society lost its functional justification. According to Simondon, twentieth-century society required horizontal relationships akin to those of the rural world. Achieving such a transformation would require, first and foremost, a reform of the educational system—one capable of providing individuals with an «intelligent learning», that is, the tools to invent new structures capables of addressing the challenges posed by an increasingly «metastable» society [5, p. 236]. The notion of *metastability*, also employed by Norbert Wiener, designates, in Simondon's thought, a state first discovered through thermodynamics. It is a state that transcends the classical opposition between stability and instability, and is charged with potential for the future. A system is said to be in a metastable state when the compatibility of its internal forces is not definitive. Therefore, the possibility remains that the equilibrium attained may be lost, prompting the system to seek a new configuration of compatibility. Simondon argues that neither the Ancients nor *Gestaltpsychology* had considered the possibility that, beyond the opposition between stable and unstable equilibrium, there could exist an intermediate condition, termed *metastable*, which he sees as the only viable solution for reconciling the discoveries of quantum physics and wave mechanics (*ILFI*, pp. 26–27, 35–36). Giovanni Carrozzini comments on the first occurrence of the term *metastable* in *ILFI* as follows:

> In the fields of chemistry and physics, metastable is defined as the state in which the substance in question exhibits energy conditions *different* from those that characterize *its stable equilibrium*. This is the state, for instance, in which «substances» find themselves during *supersaturation* or *supercooling*. Simondon defines as metastable, and thus far from stable equilibrium and *charged with potential*, the state of a system on the verge of *phase-shift*, such as, precisely, the pre-individual before the structuring of phases [35, p. 24].

Simondon proposed several strategies to achieve what he called *intelligent learning*, among which: the dissemination of information even in rural areas, which would foster a greater involvement of their inhabitants in social life without undermining the value of their local culture; the adaptation of the learning system to the adult individual's increasing capacity to learn, at least up to the age of thirty, while maintaining a constant connection between the invention of a solution and the results of the action undertaken, lest one fall into a «somnambulistic» action, utterly disconnected from inventive life [5, pp. 239–240].

In the sections entitled *Information, technologie et conditions de vie des étudiants*, one can discern the emerging connection Simondon begins to articulate between philosophy, technology, and psychology [5, pp. 239–244]. He asserted that «*consciousness must remain in phase with action*» [5, p. 244], and for this to occur, the path of an informational message must be as short as possible. In other words, within the learning process, the individual must be able to acquire a symbolic system adequate to the reality they are experiencing in order to understand the information, that is, to structure it meaningfully, without the need for multiple translations. This was an early gesture toward the idea of a «new culture», a *nouvel humanisme* capable of holding together the human and natural worlds through the technical world, which alone possesses the symbolic framework suitable for connecting the other two.

For Simondon, the technical being was the «first result of human action upon nature», and as such, it could account for both the symbolism of the maker (consciousness) and that of the user (action) through an idiographic symbolism. Technology was thus to play the role of coordinator and synthesizer of this «open» symbolism, capable of connecting multiple layers of reality without temporal disjunctions, which, if maintained, would result in profound affective and social instability for the individual [5, p. 245].

2.6 1956–57: The Proposal for a Genetic Psychology

Two years prior to the *defense* of his doctoral theses, in 1956, Simondon wrote another text entitled *Fondements de la psychologie contemporaine*, which summarized the course he had taught at the Faculty of Psychology in Poitiers [36, pp. 21–249].

Following Simondon's death, this text was incorporated into the *Sur la psychologie* collection, which was edited by his daughter Nathalie Simondon. It was accompanied by an introduction that made reference to an article that was subsequently published in 1957 by Simondon and François Le Terrier, entitled *La psychologie moderne*, and the more well-known *Forme, information et potentiels* of 1960, which was later included as a supplement to the main thesis. This enabled a more straightforward understanding of the 1956 text. In this work, he offered a critical overview of what could be defined as the three stages traversed by psychology, thereby laying the groundwork for the genetic ontology that would later underpin of *ILFI*.

According to Simondon, these stages or phases of psychology did not possess rigid boundaries, but rather pivotal centers from which one could identify the prevailing characteristics of a given phase. This was because psychology, in his view, essentially existed in a state of simultaneity between science and culture, a position that was always and inherently «dissymmetrical». On the one hand, it assimilated the results of scientific inquiry and appeared deeply influenced by them; on the other, it served as the origin point for cultural movements.

By using the two world wars as historical markers, Simondon was able to outline three distinct stages in the development of contemporary psychology: the first from 1880 to 1918; the second from 1918 to 1939; and the third beginning in 1948 (the year of the publication of *Cybernetics*). Each phase corresponds to a set of principles from which the respective psychological theory would emerge causal dualism in the first; systematic monism in the second; and genetic pluralism, or recurrent causality, in the third.

2.6.1 First Phase: Causal Dualism

The initial phase in the development of contemporary psychology would appear to be driven by the development of biological studies that had taken place in the preceding decades. Indeed, as Simondon himself recalls, biological theories established prior to 1918 only subsequently came under critical scrutiny thanks to psychology:

> By 1918, nearly all the paths we know today had already been opened; the greatest discoveries had been made. After 1918, these discoveries would undergo critical analysis, as a logic of psychology (in the same sense as one speaks of a logic of the sciences) could begin to take shape by re-examining the results already obtained. Nonetheless, the pioneering and foundational phase is indeed the one that concludes with the First World War. At this point, psychology is the daughter of biology [36, p. 25].

The latter thus takes its first steps by developing its theses in response to the biological discoveries then emerging, along with the related critiques. Structural psychology, whose founding figure is Wilhelm Wundt, indeed takes its cue from the methodology employed in biological research. Wundt established the first psychology laboratory in Leipzig, attracting scholars from across the globe who, after a period of training, returned to their home countries to found new psychology laboratories.

Wundt's research was based on the theory of the *personal equation* developed by the German astronomer Friedrich Wilhelm Bessel, which held that no observer could be entirely reliable, since the very act of observation already introduced a distortion of reality. Bessel had noted that the discrepancies in data recorded by different astronomers were not due to differences in the observed reality itself, but rather to the modifying effect of the observer upon the reality being perceived. The theory of the personal equation would go on to influence various fields of knowledge, including sociology and medicine, by casting doubt on the very act of field research, attributing to automation, previously thought to be the domain of mere machines, the sole guarantee of certainty.

Wundt also took up the research of Dutch physiologist Franciscus Donders on the subtractive method, which posited that any complex psychic process would take longer to complete than simpler ones, as reflected in reaction times. For instance, if a subject is required to respond with a single reaction to a set of stimuli, the subject will necessarily require more time to analyze all the stimuli and generate an appropriate response.

Building on these influences, Wundt's experimental theory asserts that the mental causality of *apperception*, or of passive mental processes, could be studied through the subtractive method, thanks to the notion of *the field of attention*. The mental causality of active processes, by contrast, could be understood through the laws of connection between elementary processes derived from immediate internal experience: the first of these is the law of creative synthesis, and the second, the law of psychic relations.

Wundt's theories laid the foundations for all those who would go on to inaugurate other laboratories of experimental psychology (e.g., Titchener, Bourdon, Cattell, and Spearman) and helped to disseminate the idea that mental processes should be broken down into their constituent elements in order to be analyzed through methodologies and techniques borrowed from the physiological domain.

In the same years, Wundt's method was increasingly opposed by Brentano's *psychology of the act*, which would provide the theoretical basis for *Gestalt psychology* as well as for the Würzburg School. Brentano's psychology of the act starts from the premise that psychological inquiry should focus on the act itself, as it is the act that confers a specific meaning upon an object: «which only has meaning by reference to an object, and which provides that object with its signification—thus resolving the associationist problem of the meaning of perceived things» [36, p. 30]. At the same time, the act acquires meaning only in relation to the object. As a result, there emerges an inseparable relationship between act and object, which makes it impossible to study the one without the other.

As early as 1874, Brentano's project was to establish an empirical, scientific psychology grounded in Positivism. In particular, in *Psychology from an Empirical Standpoint* [37], Brentano clarifies what he means by perception and how it occurs. His thesis was based on the intent to found what would later be called a *psychology without a soul*, a psychology whose goal was no longer the study of the soul or psyche, but the analysis of that which can be experienced [36, pp. 30–31]. Within this broader domain, Brentano identified two types of phenomena: *psychical phenomena*, which he called «presentations», and *physical phenomena*, which concern objects external to the subject and are unrelated to the subject's consciousness. Only these two types of phenomena can be experienced in any meaningful way, and only they can become the objects of this new empirical psychology.

It was Brentano's philosophy that inspired Ward's proposal in England to open a psychology laboratory that could oppose associationism. Despite the efforts of Ward and his students to make psychology a science worthy of being studied like any other, it did not enter British academic studies until after the First World War. Following Brentano's thought and remaining faithful to a biologist view, Ward argued that the mind cannot be a collection of sensations but, rather, must be studied and thought of as mental life in which activities take place [36, pp. 32–33].

In France, experimental psychology influenced by biology is mainly headed by Henri Piéron and Marcel Foucault [36, p. 38]. The latter defended a thesis on psychophysics in 1901 and founded a laboratory for analytical psychology in Montpellier. In 1911, Piéron became the director of the Sorbonne's psychology laboratory and the attached journal *Année psychologique*, through which French scholars were able to get in touch with the research that was taking place in the rest of the world. The theory pursued by Piéron is deeply imbued with biology, especially through the link woven with physiology and evolutionary theory [36, p. 39]. Every anatomical structure is explained from the development of the function it performs. Simondon highlights how important the biologism revolving around the experimental psychology of Piéron and his pupils was because not only did it make it possible to overcome behaviourism by giving ample space to the theory of evolution, but also «it incorporates concepts that seem to belong to theories that appeared later, such as the 'pattern' or organismic unit. It prefigures certain aspects of information theory (notion of the perfect receiver; background noise)» [36, p. 41].

The influence of biology on psychology is decisive not only for experimental psychology as expressed in its various theories (German, French, English, American and Russian), but also for theoretical psychology, in particular that which emerged from the texts of Henri Bergson [36, p. 52]. Although Bergson's vitalist theory is opposed to the extreme analyticity of studies in experimental psychology, it shares with the latter, according to Simondon, a profound influence from the life sciences. Two souls coexist in Bergson, that of vitalist monism and that of psycho-physiological dualism, which are reflected in his theory of action. According to Bergson, action can be considered both as work and as invention: it has a «double statut» [36, p. 53]. In the former case, it is a utilitarian view of action that binds man to materiality, makes him *homo faber* and builds his habits. As invention, on the other hand, action is thought of as *élan*, a free impetus completely disengaged from the patterns of adaptation (which Bergson sharply criticises). In spite of the dualism that emerges from this theory of action [36, p. 54], Bergson places man within a vital flux on which his gaze becomes monist again:

> He was profoundly monistic, studying man from the perspective of life; the very oscillation between dualism and monism in his work merely reflects the evolution of psychology since the end of the 19th century. [36, p. 57]

In the context of psychology, Bergsonian thought is characterised by a decentralisation of the quantitative conception of the psyche and states of consciousness. This is achieved through the assertion that the pure psyche, the deep self, concealed by the superficial self, is a spontaneous flow of duration. The only means of apprehending the pure psyche, it is asserted, is through intuition, which the psychologist would only attain following a profound transformation. The resulting psychological knowledge would be considered metaphysical in nature, as it would facilitate the comprehension of the psyche in a single moment and, concomitantly, enable participation in the dynamism of the pure psyche. This dynamism would be expressed by the spontaneous flow of duration within the subject [36, p. 56].

However, what appears to be pure monism still conceals a dualism that Simondon captures in *Creative Evolution* [38], in which the notion of function is ambivalent. On the one hand, it is the driving force of life, the highest expression of life from which evolution has developed and continues to develop,on the other hand, there remains the materialist aspect that perceives function as adaptation, constraint or obligation to a life «stopped by the matter» [36, p. 55]. Simondon hypothesises that this ambivalence, coupled with a pronounced resistance to scientism, may have contributed to Bergson's exclusion from the realm of psychology. This «concealed dualism», whereby life is both the static nature of adaptation and the dynamism of *élan*, would have led Bergson to be taken up in Norbert Wiener's *Cybernetics* in the chapter on the comparison between Newtonian and Bergsonian time [36, p. 56]. Bergson's conception of psychic reality as both *molar* and holistic would have provided the intellectual foundation for challenging the late 19th-century scientific determinism that dominated the field, thereby opening up new avenues for research in psychology and related disciplines. In psychology, the *molar conception of the psyche* refers to an approach that considers behaviour as a global and meaningful unit, which cannot be broken down into individual elements or muscle movements. Instead of analysing behaviour by breaking it down into small molecular units (such as individual movements), the molar approach studies it in its entirety, taking into account factors such as intentions, purposes and meaning. An example could be the following: an individual washing their hands. A *molecular approach* could describe the movements of the hand, but a *molar approach* would consider the gesture as part of a broader action, such as personal hygiene or disease prevention, taking into account intentions and context.

2.6.2 Second Phase: Systematic Monism

The second phase, called *systematic monism* by Simondon and articulated in the years between 1918 and 1939, is the one in which psychic reality begins to be thought of as a whole within an organism, overcoming the dualisms established in the previous phase: «psychic reality is grasped as the whole of the behaviour of an organism adapted to the environment of the molar reality of a whole with a form» [36, p. 87].

It can be defined as a phase of settlement as far as discovery is concerned, but also one of profound revision of what had been discovered.

Watson's *behaviourism* began to assert itself in the United States in response to the opposition between structuralists and experimentalists [36, p. 88]. The object of study of behaviourism is no longer the action of consciousness on the body, as it had been for the psychological theories of the first phase, but behaviour, i.e. that dynamic enacted by the subject in order to seek equilibrium with respect to changes in the surrounding environment. Until then, behaviour had been studied on animals and not on human beings. The motivation lies in the fact that studying behaviour means, in fact, reducing the action of consciousness to a stimulus–response dynamic: the situation in which the subject finds himself brings within it a series of stimuli that act on the organism, which attempts to respond to them through the activation of certain parts of the body. Watson proposed to eliminate this assumption and extend the study of behaviour to the human being [39].

Behaviourism can be regarded as an initial monist response to the dualism imposed by the first phase, which had failed to identify a common ground between psychological theory and the introspective method. Watson thus proposed the abandonment of the introspective method, which he deemed incapable of studying behaviour, and instead advocated a shift towards case studies, which enabled the emergence of pure behaviour as stimulus–response dynamics.

Watson's proposal began to circulate in the United States and Europe as early as his first article in 1908. In France, behaviourism was immediately embraced by Piéron, who welcomed the new approach. Through correspondence with Simondon, who sought access to the Sorbonne laboratories, the two scholars exchanged ideas on the subject.

Concurrently, in Europe, particularly in Germany, a psychological theory was evolving that would prove to be a pivotal nexus for certain concepts that Simondon would subsequently appropriate in an attempt to integrate psychology and philosophy. The theory, known as *Gestaltpsychologie* or Form Psychology, was pioneered by Max Wertheimer, a student of the structuralist Oswald Külpe and Carl Stumpf, a proponent of Brentano's theory. *Gestaltpsychologie* represents the unification of philosophical and analytical psychology, which emerged towards the close of the nineteenth century. The concept of the field, derived from physics, served as the unifying principle. The philosophical underpinnings of *Gestaltpsychologie* are predicated on the notion of form (archetypal or instinctual) while the analytical school of associationism emphasised the distinction between content and act. The concept of field, however, served to synthesise these seemingly contradictory dualisms:

> Compared to these two ancient notions of form, the psychology of form constitutes a notional discovery, because the form-structure of Gestalt theory combines the fundamental characteristics of Platonic form and Aristotelian form, thanks to the discovery of a type of constitutive relationship that neither Plato nor Aristotle knew about, and whose model is provided to theorists of form by the Physics of the early 20th century: the community of action in a homogeneous field [36, p. 91].

Gestaltpsychologie is predicated on the notion that there is no mechanical or genetic randomness in the psychic field. According to this thesis, every psychological phenomenon is posited to be nothing more than the set of self-structuring characteristics of an autonomous field—that is to say, a system that, in order to structure itself, does not refer to the past, but only to the present [36, p. 91]. According to Simondon, this implies that all things are immanent in the present, and that there are no latent potentialities that can be actualised.[6] The concept of a field facilitates the integration of content and collaborative action, which is structured within it according to precise rules. A secondary issue of *Gestaltpsychologie* pertains to the fundamental rule of structuring a field, that of good form, which was introduced by von Ehrenfels' acoustical studies. According to this rule, every field tends spontaneously towards the most balanced, most homogeneous state. The concept of good form signifies the tendency of a field to gravitate towards a state of optimal balance and homogeneity, through the process of attempting to organise any imbalance that arises from the disparity present within two distinct regions of the given field or system [36, pp. 93–99]. In the context of psychology, the concept of good form pertains to the segregation of perceptual units in accordance with a propensity towards the closest or analogous stable equilibrium:

> This law is given as capable of accounting for an appearance of finality which is in fact only a result of causality in the form of the community of action of the elements of a system: order presupposes no principle higher than the reality of the domain in which it exists, no intervention by anything other than pure causality. Order is opposed to disorder in the same way as the domain of forms is opposed to that of additive groupings, of strong forms to that of weak forms [36, p. 97].

This notion of good form would be strongly criticised by Simondon, especially in *ILFI* and in the 1960 article, as being the weak point of *Gestaltpsychologie*. Although they introduced the concept of the field by opening up the possibility of a thesis on energetic potentials, with the concept of good form the first-generation Gestaltists would foreclose this possibility by maintaining the opposition between stable and unstable. Furthermore, they claimed that the best form/structure/organisation is that which does not involve further change. In fact, following the laws of thermodynamics, *Gestaltpsychologie* argues that the good form of a system is that which does not imply possible further transformations, thus that system in which entropy would be at its maximum degree.

In Simondon's historical reconstruction, one of the major evolutions of *Gestaltpsychologie* occupies a prominent place, given by the encounter with American-style behaviourism, namely Kurt Lewin's dynamic theory of form or odological theory. The latter, on the one hand, accords the concept of organisation greater prominence than the notion of a field in stable equilibrium and, on the other, arrives at the idea that the individual being is a part of a whole made up of it and the environment. This whole is organised or structured by different sub-fields such as the Ego and the environment that stand in asymmetrical relationships with each other: «an asymmetrical relationship then emerges between the environment and the ego, as between the figure and the background in Wertheimer's experiments on perception» [36, p. 102]. Form,

[6] As will be seen later, the issue of the absence of virtuality will be one of Simondon's critical points.

then, is this asymmetrical relationship that, as such, abandons the aura of stability by which it was cloaked in classical *Gestaltpsychologie*. Lewin is driven towards this evolution not only by his encounter with behaviourism, but also by discoveries in the field of physics (Bohr's quantum theory and Heisenberg's uncertainty principle) that were undermining the certainties of late nineteenth century physics on which field theory was also based. In particular, both Bohr's theory and Heisenberg's principle inject into the field of reality an energetic discontinuity between what is visible and what, despite not being visible, has a profound effect on that reality. Transposing Bohr's and Heisenberg's findings into the psychological field, an individual's behaviour would thus become the result of a dynamic field in which various forces act and collide [36, pp. 102–103].

According to Lewin, psychology would continue to be constrained by the Aristotelian conception that subdivides psychic life into separate fields, without considering the potential for interaction between these domains. The psychological conception proposed by Lewin sets itself the task of searching for general laws that demonstrate the interactions between psychic fields, which, therefore, cannot be separated from one another.

For Simondon, the psychology of form represents a pivotal nexus in psychical studies, as well as in the establishment of psychology as a scientific discipline. The Gestaltist method pioneered by Simondon entailed a dynamic interplay between the principles underpinning classical psychology and the emerging findings in other scientific domains, including physics. It was precisely the Gestaltist method that Simondon inherited to elaborate his psychological theory within *ILFI* and *MEOT*. In particular, within the Lewinian phase, Simondon acknowledges an overcoming that he deems essential to his ontology, that of isomorphism. The latter proved fundamental to the initial phase of *Gestaltpsychologie*, which posited that the physical, physiological and psychic domains possessed a uniform structure.

A more pronounced departure from isomorphism is represented, according to Simondon, by philosophical psychology or phenomenological psychology. Husserl's critique of psychologism can be seen as the foundation for three distinct phenomenological currents: Goldstein's, Sartre's and Merleau-Ponty's [36, pp. 113–114]. A further distinction between these approaches lies in their relationship to *Gestaltpsychologie*, which is outlined in the following section:

> It does not seek to constitute an encyclopaedic unity of knowledge, but accepts from the outset a specific difference between the different ways of reaching the object, relegating the scientific way to the inessential and opposing it to the philosophical way, which is a search for essences by eidetic reduction [36, pp. 113–114].

In Goldstein, Simondon traces the privilege accorded to the notion of organisation found in Lewin, but also the complete detachment from the concept of good form as the stable equilibrium of a system. Furthermore, in Goldstein's phenomenology, the relationship between organism and environment is central in that every manifestation of the organism takes on meaning only within the environment. Reflexes, for instance, are posited as a reaction of the organism aimed at finding a compromise with the environment; without this justification, the reflex would be devoid of meaning.

According to Goldstein's theory, the environment functions as the organism's counterpart, a prerequisite for the emergence of culture as the nexus where the drives for novelty and the tendency to structure in order to maintain order intersect. This process, as posited by Goldstein, occurs through the collision of the organism's forces of novelty with the environment, in pursuit of establishing order and potential adaptation. Goldstein's approach diverges from causal thinking as embodied by *Gestaltpsychologie*, instead introducing the concept of non-causality into his «biological» phenomenology, drawing upon Bohr's quantum theory. Goldstein's approach involves distinguishing between continuous and knowable states, on the one hand, and discontinuous and incomprehensible states, on the other. He contends that the former are observable in the macroscopic order of reality, while the latter are observable in the microscopic order. The influence of the latter on the intelligible, macroscopic real is a key tenet of the theory. Psychological reality is part of the microscopic order, the discontinuous one, and so cannot be explained causally except by isolating individual phenomena [36, pp. 115–120].

In contrast, Sartre's approach, drawing upon the Husserlian concept of eidetic intuition, posits that the transcendence of psychologism is achievable only through an examination of the fundamental essence of the human condition. Psychic facts are defined as man's reactions to the world and are tasked with the superimposition of these two dimensions. The transcendental consciousness facilitates the placement of the world «in brackets», thereby enabling the discernment of the essence of the human being.

By employing this methodology, Sartre is able to examine emotions as pure transcendental phenomena, which facilitate the connection between man and his essence. The dialectic between emotion and consciousness is characterised by a reciprocal relationship, with the former serving as a medium for the articulation of the latter. According to Sartre's interpretation of Heidegger,[7] human essence is defined by the assumption of self within an existential world of understanding, and the capacity to be moved is identified as the fundamental characteristic of the human condition. Accordingly, consciousness is defined as the necessity of emotion itself. Sartre's approach diverges from psychologism in its conception of emotion as an organised system of means to an end, rather than merely a response to external stimuli.

In Sartre's phenomenological psychology, the Gestaltist concept—which posits that consciousness orchestrates transitions in form and diverse transformations through purposeful synthetic action—and the Freudian thesis—which asserts that each state of consciousness possesses distinct value—coexist. At the same time, the identification between signified and signifier remains, which entails the idea that the meaning of a behaviour is internal to the behaviour itself and that consciousness does not consist of several parts, but already contains the meaning [36, pp. 129–131].

[7] This is Simondon's second reference to Heidegger. In this case as a mere quotation from Sartre. On the other hand, in *MEOT* Simondon thanks Professor Canguilhem for giving him access to some also very rare German texts. Therefore, it is possible that Simondon had access to Heidegger's texts, but he never quotes them directly.

With respect to Merleau-Ponty's phenomenological psychology, one of the approaches that would most inspire Simondon's theory, it is important to bring this course into dialogue with an article Simondon wrote together with Francois Le Terrier in 1957 and entitled *La psychologie moderne*. Published within the volume dedicated to science in the *Encyclopédie de la Pléiade* edited by Mauroce Daumas [40], in it, unlike illustrious colleagues such as Fraisse, Piéron and Fessard, Simondon argues for the need to trace the common purpose of psychology and philosophy to questions concerning human nature. A disconnect between the two disciplines in order to make psychology more «scientific» would only miss the very questions of psychology. In this sense, Simondon's is a strenuous defence of Renaissance culture (and subsequently encyclopaedic culture) that had been completely lost in the extreme specialisation demanded by scientism and that, instead, he considers fundamental if we are to understand reality as an organic and complex whole. In particular, Simondon argues that the phenomenology of Husserl and Merleau-Ponty would allow the problem of psychology to be refocused from the eidetic presupposition that assigns to consciousness the role of elaborating the meaning of facts before the facts themselves can be arrived at. In particular, Merleau-Ponty is made the object of attention insofar as he takes up the later Husserl and attempts to further explain the concept of *Lebenswelt* through the Gestaltist idea of the *Umwelt* while excluding the causal derivation of form theory [40, pp. 1678–1679].

Merleau-Ponty rejects causal thinking as merely descriptive and not explanatory, arguing that the human being is a subject devoted to the world, to that *Lebenswelt* which represents the field in which all his thoughts and perceptions are present. It is a field that precedes any analysis that the subject can make, and in this sense perception is the background against which all acts take place. The subject is able to distinguish itself from the world and identify its body, which it is able to separate from itself, as a thing among things. This can happen thanks to the cogitation that grasps the human being always in a situation and the fact that transcendental subjectivity, as Husserl wanted, would always be an intersubjectivity [41, 90]. In this passage, Simondon seems to attribute the concept of *intersubjectivity* to Merleau-Ponty, effectively missing the necessary passage from Husserl. It must be said that, as Husserl scholars themselves point out, the centrality of the concept of intersubjectivity in his thought has only been discovered since the publication of the unpublished *Gesammelte Werke*. This could mean that Simondon, who possibly did not have direct access to Husserl's texts, relied on the references in Merleau-Ponty's texts and assigned him authorship. In fact, the latter presents intersubjectivity as early as *Ideen zu einer reinen Phänomenologie und phänomenologischen Philosophie* in 1913 and then takes it up again in the *Mèditations cartèsiennes* of 1929–1931. Husserl's aim is to ward off the solipsistic drift on the part of the transcendental ego even though he considers the solipsistic illusion that emerges from the phenomenological epochè to be natural. In empathy Husserlfinds the way through which the subject experiences the existence of other subjects and at the same time manages to overcome his subjective view of the world (*Erlebnis*) in order to arrive at an objective communitarian view (*Lebenswelt*) that allows access to essences. Husserl defines intersubjectivity as a «transcendental monadic community» that is established in the second moment of transcendental phenomenology, the truly philosophical one, in which «transcendental subjectivity» is no longer meaningful only for the *Ego* but also for other subjects.

For Merleau-Ponty, the world is what the subject perceives, and intentionality (another pivotal concept taken up by Brentano through Husserl) is not only what allows consciousness to recognise itself as a project of the world, but also what reconstitutes, as operating, the natural and pre-predicative unity of the world itself. Merleau-Ponty thus aims to extend the Husserlian conception of intentionality by accentuating its intersubjective dimension, a total intention to which man arrives from perception and understanding of the world [36, pp. 132–134]. In Merleau-Ponty Simondon finds what may be called a phenomenology of genesis, to which he owes much for his idea of a genetic ontology.

In contrast, the Merleau-Ponty vision aims to extend phenomenology beyond its existing limits, both as a biological phenomenology and as a phenomenology of genesis. The question of the body is central to this aim. The theme of corporeality was already present in Husserl's distinction between the *Leib*, i.e. the living body connected to the psyche, and the *Körper*, i.e. the object body that is not endowed with a connection to the psychic world and, therefore, cannot be considered living. In *Ideen zu einer reinen Phänomenologie und phänomenologischen Philosophie* (book two) Husserl specifies that the body endowed with life is the «perceptual organ of the experiential subject», the phenomenological subject that possesses the transcendental life of consciousness. Furthermore, the Husserlian Leib lives in the Leiblichkeit, in the corporeity that distinguishes it both as a human subject that feels its body and as an individual in the uniqueness of its corporeity [43]. Merleau-Ponty would take up the theme of corporeality and make it central to his phenomenology [12, 44].

The body is instrumental in facilitating perception and comprehension of the world, serving as the medium for being-in-the-world, a concept within Merleau-Ponty's philosophy that encompasses the unification of the psychic and physiological dimensions. This unification is conceptualised through the notion that the physical world is characterised by specific stimuli and situations (personal existence) and an *impersonal* existence that surrounds it. The latter is represented by the organism as never single but always plural and, for this reason, impersonal and always present. Personal existence, it is argued, intermittently intervenes in this impersonal flux, and the body is that which maintains the link with the *pre-objective impersonal*, in which psychic and physiological are two sides of the same coin. The problem of the *impersonal* arises in Merleau-Ponty in the last phase of his reflection in clear opposition to the initial assonance between French phenomenology and personalism. In a radical Husserlian post-phenomenology, the Merleau-Ponty of the 1950s above all wants to move towards a 'new ontology' marked by the profound relation between the 'carnal', pre-individual and pre-personal or impersonal, of entities. Shifting his gaze to a perspective that can be said to be non-anthropocentric and markedly anti-dualist, Merleau-Ponty argues that every movement in the world of the individual is grafted into a tangle of relations that interact with the individual itself, and that subjectivity or consciousness can only emerge from an original layer that precedes the structuring of the individual. One is no more aware of being the true subject of a sensation than one is aware of one's own birth or death, according to Merleau-Ponty, and this means that the sensation is rooted in a prepersonal field that survives individual existence. For a more in-depth study ofthe link between Merleau-Ponty's impersonal and Simondon's pre-individual [45], 46, pp. 119–132, 47, pp. 393–409].

This theoretical framework enables Merleau-Ponty's conceptualisation of diverse forms of behaviour—including the sexual, cognitive, and action-oriented—as components of a unified structural whole. This notion has also been traced back to psychoanalysis, which reintegrated the libido within the human being as that which first passes through consciousness and then through the body, forming a whole with the human being itself.

A further author marking the second phase in the development of psychology is the Swiss psychiatrist and psychoanalyst Carl Gustav Jung, who attempted to make a synthesis between Freud's and Adler's theories through his «psychology of the deep»: «he notes that there are two psychic attitudes in man, introversion and extraversion, orienting the ego towards the inner world or the outer world, and that each forms the basis of a type of neurosis» [36, p. 137]. The concept of the «complex» is the key to understanding Jung's particular position that will influence Simondonian thought. The complex is a structured set of representations that may be conscious or unconscious but always endowed with a certain emotional/affective charge. The human psyche is made up of several complexes,of these, those endowed with greater power are the ego complex and what Jung calls the *collective psyche*: the dimension from which come the archetypes acquired during the evolutionary process of the species. These archetypes emerge during dreams and are the same in all human beings. Within the nervous system, it is the *libido* that through which, for Jung, the energies of the psyche are expressed. This activity consists in holding the personal ego complex in relation to other egos. When this relationship begins to lose strength, the ego loses control of the other complexes, which begin to struggle to take over, giving rise to neurosis. The lack of communication between the Ego and the other complexes leads, therefore, to dissociation because the human being is unable to recognise in the messages received the presence of libidinal drives (which are not only equivalent to sexuality, but to all psychic energy) that open up the possibility of transformation. In Jung, psychic life is not anchored to the past as in Freud but is projected into the future thanks to the production of symbols that allow the oppositions present between the complexes to be overcome. The process that makes it possible to overcome the opposites that constitute the psyche of the human being through division and integration is called individuation in Jung. It is a process that allows the ego to stabilise itself by interacting with the unconscious and organising the energies present in it by adapting them to reality through participation in the collective psyche [36, p. 138].

The first hint of what Simondon properly calls a genetic psychology in the third phase of the development of contemporary psychology can be found in the child psychology that emerged after 1914. If in the first phase child psychology was seen as a new domain of psychological research that did not interact with the other domains, in this second phase the study of the child became a method, a specific way of doing research on the human being. Still steeped in behaviourism, the first exponent of child psychology from a monist perspective was Leonard Carmichael [36, p. 167]. Beginning with the study of behaviour in the early stages of human development, Carmichael sought to discover its origin through interaction with embryology. It is this research that opens the way towards solving much broader psychological problems, such as the connection between the development of the mental life of an adult human being, genetics (heredity), and epigenetics (environment). Carmichael stated that the study of prenatal behaviour could shed light on the issue of learning, and that by abandoning introspection and simple observation, child psychology would make it possible to investigate areas that had hitherto remained in the shadows, as it would give the possibility of studying different behaviours at the moment of their genesis. Only with a genetic method, in fact, can later developments be understood.

The only limitation found by Simondon in this proposal is a behaviourist assumption that translates into the claim that all motor organs, muscles and nerves must be sufficiently developed for one to be able to speak of behaviour, i.e. reaction to an arousal. It is well known, however, that in the gestation process of an infant not all these parts develop at the same time and at the same rate. Therefore, to speak of prenatal behaviour in the behaviourist sense would lead to a short-circuiting of the theses advocated.

The psychologist that Simondon will focus on most in his later work is Arnold Lucius Gesell [36, pp. 170–172]. Accepting Carmichael's premises, Gesell devoted his studies to demonstrating the complete structural analogy between animal psychology and child psychology through physiological studies of experimental embryology. These studies in fact demonstrate that somatic development corresponds to psychic development: in both cases it is a matter of morphogenesis, i.e. the progressive generation of forms.

Studies on morphogenesis, however, were still very meagre at the time, and Gesell could only arrive at a definition of certain principles of development that he claimed would have consequences in both the somatic and psychic domains. By defining these principles it should be possible to study, for example, the genesis of motor behaviour in which a continuity between the prenatal and postnatal phases can be inferred.[8]

2.6.3 *Third Phase: Genetic Pluralism*

The development of child psychology in the second phase of the history of contemporary psychology serves as a foundation for the third phase, explored during the years in which Simondon lectures at Poitiers. In the initial phase, the nexus between psychology and other sciences was predicated on the findings in the pursuit of a

[8] Cf. https://web.archive.org/web/20180313094130/http://childstudycenter.yale.edu/about/history.aspx.

principle of causality. In the subsequent phase, this association was founded on principles endeavouring to ascertain a general axiomatics. In the third phase, however, a paradigm shift emerges, where the nexus is posited to reside in an intermediary phase between principles and results. The theoretical framework underpinning this psychological development is that of interference, a concept that facilitates the observation of a dynamic process without the need for a priori principles or predetermined outcomes. Indeed, scholars within this third phase seek to establish an analogy between psychic and life processes without presupposing an absolute identity between life and thought [36, p. 181]. This phase of research, termed "genetic pluralism" by Simondon, is predicated on the identity of processes rather than substances. In his course of 1956, Simondon's first stance emerges with regard to the need to change perspective on the question of structural analogy that had, in fact, also affected the psychological study of the previous phases in an attempt to bring it back to an identity with the somatic:

> Functional analogies are of great value for research, and allow the foundation of fields of study that are not based on a classification of types of being (matter, life, spirit), but on types of dynamism of beings [36, pp. 181–182].

In order to speak of functional analogy and no longer structural analogy, it is necessary, in the first instance, to abandon the search for the causes of phenomena and to focus on effects. Simondon brings his scientific studies into play and uses the concept of effect in the sense in which it is used in physics: a complex of phenomena and operations, a node of organised operations that occurs under certain conditions independently of the experimenter. The effect is distinguished from the result because it has its own internal stability and has the capacity to organise the system in which it is triggered, i.e. when a certain threshold is crossed that generates it.

Therefore, the effect is neither a phenomenon nor a result, but the process that occurs between these two poles. According to Simondon, psychology, in its genetic phase, must concern itself with investigating this third form of the relationship which, as will be made clear later in *ILFI*, is a condition of the relationship itself. Psychological research must not be driven by the need to find a cause and a result of a process before it takes place, thus assuming that the process will unfold in a certain way.

In order to frame more clearly the direction that psychology should take, Simondon appeals to technology and its being, par excellence, the investigation of effects. Simondon had already read [48] text and confronted cybernetics in his 1953 essays. It was now very clear to him how the new technology, that of information technology and statistics, was the link between the various particular sciences that could no longer think of themselves as truly independent. Technology thus became the study of the operations between these sciences and no longer a mere application of the laws discovered by them. Simondon agrees that, from the earliest times, techniques and the construction of technical objects have come before explanations of the laws that governed them [49].

Norbert Wiener was the father of this encounter between psychology and technology in what he called *Cybernetics*, or the study of control and communication in humans and animals [48]. Cybernetics could become the keystone of psychology not only because it promises to earn it that scientific status that scholars yearned for, but because it does so without coming to terms with the tendencies of the first and second phases of its development. It has been said that in the first phase what drove psychological research was the need to find the efficient causality of *phenomena*; whereas in the second phase the focus was on the results of processes triggered by the efficient cause. With cybernetics, the process-oriented study carried out by Wiener's mathematical analysis points to a circular causality, i.e. a process in which the causality that is triggered from the effect to the effector is of a different type than that which operates from the effector to the effect. To clarify further: in a cycle, there is a phase in which there is a taking of information which, through the effect, allows the efficient cause to transform itself into the conditioning cause; and then a phase in which the latter, through a process of modulation of the energy present in the system, controls the efficient cause. This second type of causality cannot cause the effect, but can only modulate it, i.e. organise, control, a certain amount of energy present in the system without adding more.

Cybernetics is therefore concerned with studying circular causality and information as essential features of any type of effect, i.e. of any structure in operation. Transposed to the psychological level, this implies a study of the psyche that follows the same essential routes, as what we call psyche is nothing more than a structure in function, an effect from a process of circular causality in which there is energy/information exchange. The psyche ceases to represent both a dead structure whose origin or efficient cause must be traced and a structure whose behaviours or effects must be studied.

One of Wiener's best-known types of research on the psychological issue and reported in the 1948 text is the one concerning the prediction of the trajectory of warplanes' aiming at the same time as solving the problem found in calculating machines when faced with functions with several variables [48, pp. 71 and ff.]. Wiener's proposal, assisted by the engineer Julian Bigelow, is to consider man and the calculating machine in the aircraft as a system in which the two parts influence each other: one must take into account the programme entered into the machine, the hand that requires the calculation and the fact that the aircraft from which the shot will come is driven by a pilot who will have his own specific driving characteristics. The key term used by Wiener to solve this problem is *feedback*: a process in which there must be a part that can act as an information signal, as *input* (*feed*) and another part that indicates a certain recursiveness of the *input information* (*back*). In this sense, Wiener's strongest thesis is that feedback, as a recursive information dynamic, is not only present in machines or human–machine systems but must also operate in human–human relations and in the functioning of the psyche. Indeed, Wiener's text opens with a description of the results obtained in subjects in which the feedback system was missing. When this system lost its capacity, brain disorders occurred that led to motor problems. The brain, therefore, cannot be considered an isolated system, but a totality of feedback interactions [48, pp. 133 and ff.].

Simondon retraces the history of the birth of cybernetics by showing his students how fundamental the contribution of Wiener and his research group was to the birth of a new psychology. No longer dedicated solely to the problem of form perception and the search for common notions through which to encapsulate perceptions, psychology became, in Wiener and the cybernetics community, a study of the nervous system in continuous exchange with sociology, anthropology, information theories and electronics. Wiener included several psychologists in his group of cyberneticians, and the meetings held in New York at the Josiah Macy Foundation from 1946 onwards were attended by several nervous system scholars and psychologists, including Kurt Lewin himself [36, p. 193].

The influence of the Macy Conferences group is so impactful that Fessard, as early as 1946, included cybernetic theories on the analogy between nervous system operations and electronic model operations in his neurophysiology courses. Simondon had taken Fessard's courses, and it is clear where the inspiration for many of the theses he later expounded in his work on the question of functional analogy came from [36, p. 194].

Simondon emphasises the presence in France of perspectives opened up by cybernetics as early as the 1930s, with the theses of Louis Couffignal [36, pp. 196–197] and the moving away of French theorists from American cybernetics to inaugurate their own original strand [50]. Couffignal's work dealt in particular with the comparison between human functioning and the functioning of calculating machines as early as 1933 with the text *Les Machines à calculer, leurs principes, leur évolution* [51] and then in *L'analyse mécanique* of 1938, *Les machines à calculer* of 1942, and *Les Machines à penser* of 1952. Unlike Wiener, Couffignal concentrated his studies on analogue machines and demonstrated the reciprocal influence that machines and human calculating requirements had in their development. Couffignal thought of comparative mechanics as the specific element in the French perspective on cybernetics that was able to overcome the great limitation of the American alternative: its reliance on the method of analogy between structures, rather than the «analogie entre fonctions» that Couffignal proposed instead [50, pp. 237–238]. He criticised American cybernetics in that it had attempted to structure a theory of machines based on Turing's idea of the universal machine, whereas according to the French scientist the machines to be considered should be those «construites jusqu'à présent», moving away from any kind of mental experiment. This, in fact, would have meant building virtual machines and risking forgetting that «l'accident est fréquent, si l'on part des mathématiques» and getting stuck in a delusion like the one caused by Maxwell's demon [52, pp. 9, 46–54, 336–337].

Simondon recalls that cybernetics had arrived in France as a theory of signals and information thanks to the figure of Louis De Broglie who, as early as 1950, had organised meetings to bring Wiener's theories into dialogue with those on information and entropy elaborated since 1925 by Szilard and taken up by Brillouin:

> Louis de Broglie traces the first studies on information theory back to 1925, with Szilard's work on entropy and Maxwell's demon, which was taken up by Brillouin [36, p. 197].

Around De Broglie had gathered many scholars who, following the interdisciplinary spirit of cybernetics, had developed studies that would hold together the purely technical as well as the psychosocial issues that Wiener himself had dealt with.

Meanwhile, in the United States, meetings at the Josiah Macy Foundation were giving rise to important collaborations that would increasingly accentuate the psychological aspect of cybernetics research: «the 1949 symposium, entitled Conference on Cybernetics, continued the research undertaken by Norbert Wiener, emphasising the psychological scope of the work, in close connection with advances in neurophysiological research» [36, p. 198]. One of the most heated debates would be the comparison between the characteristics of the knowledge acquired by humans in the "classical" way (i.e. from other humans) and the same knowledge acquired by a machine. To prevent knowledge from being compromised, Wiener argues that there must be as little resonance as possible between the measuring/knowledge instrument and the object to be known. Several scholars proposed, then, to train machines in such a way that they have within them the tools necessary to know a subject (the only human in the relationship) without the intervention of another human being. Basically, the machine would have to be programmed in a completely automatic way, even with respect to the questions to be asked of the subject to be known/measured, in order to exclude any possible indeterminate event [36, pp. 199–200].

A counterproposal came from child psychologist Margaret Mead, who argued that it is not possible to know everything about the human being through this method because man cannot be compared to a machine among machines [36, p. 201]. In fact, Mead argued that it is impossible to establish a structural analogy between machines and humans and that the resonance that Wiener would have minimised in order to avoid distortions of knowledge actually only needs to be controlled, 'organised' by humans as a bridge between machines. One can see in this position one of the strongest Simondonian theses, dealt with extensively in *MEOT*—that of man as interpreter and co-ordinator of machines (pp. 12–13).

In this sense, Simondon criticises the French misinterpretation of cybernetics as not taking into account the fact that not all cybernetic research is based on a mere structural assimilation of man to machine. Although there are scholars who approach research following this principle, in reality, Simondon argues, «this community, transcending both the humanity of man and the automatism of the machine, which is the object of cybernetics» [36, p. 211]. Within cybernetics, there is as much a humanisation of the machine as a mechanisation of man, as there is a functional continuity between man and machine that allows permanent translation between the two individuals. Simondon points out how this idea was developed in the United States because, unlike Europe, the US is a place where the relationship with the machine is an integral part of man's life; the machine is, for the American person, a life companion; whereas in Europe, thinking about the machine is rather influenced by Cartesian mechanicism and some presupposed primacy of theoretical knowledge over technical knowledge:

> In cybernetics there is a profound faith in the destiny of man as he embarks on the mechanopoietic and technological adventure; cybernetics is a philosophy of the act, of man as producer, who accepts this reverberation of the forces of his existence that are the machines, and who assumes them instead of denying them as a necessary evil to which he has resorted because he is pressed by necessity; a certain bad faith essential to the anti-technical attitude is absent in cybernetics [36, p. 212].

Simondon closes his excursus on cybernetics with words that sound anticipatory of the critique with which he will open *MEOT* (pp. 9–16): in order to reconsider the existence value of the technical object, it is necessary to abandon the European cultural prejudice according to which machines are only tools created for certain needs and to be kept under control because they are potentially dangerous. According to Simondon, American cybernetics overcame this obstacle precisely because it emerged in an environment where anti-technical attitudes had not become established.

Simondon had already expressed his positive opinion of cybernetics, albeit with some reservations, in his 1953 essays. These texts show how cybernetics is a fundamental starting point for his research. It is not a classical science, but an attempt to value the relationships between the sciences and to provide a place for threshold disciplines such as psychology and sociology. Cybernetics is, for Simondon, the inspiration needed to abandon a merely substantialist perspective on reality and open up a first clearing for an operational study. In spite of the encyclopaedic and operative vocation of such «no-man's land», Simondon had already pointed out its shortcomings in '53, the most important of which was the absence of a logic that could have transcended individual cybernetics (neurophysiology, information theory, etc.) and founded a universal cybernetics, called *allagmatics* (*ILFI*, pp. 529–536), i.e. a study of operations through a common symbology. Only through the transition from Wienerian cybernetics to allagmatics would it be possible to affirm the functional analogy present in the different levels of reality.

The search for a correct perspective to bring out this profound analogical link is for Simondon the impetus for the work carried out in his two doctoral theses. Only if one starts from the eradication of the substantialist presupposition and the need to search for a proper place for contemporary psychological research can one understand the intensity of the relationship between the two theses. For the philosopher, the study of the genesis and evolution of the technical object is the key to understanding the different levels of reality and a mirror in which to observe and try to clarify the functioning of the psychic level.

2.7 The *Change of Focus* with Georges Canguilhem

In his biography, Nathalie Simondon quotes a few lines from a letter written by his father in 1952, addressed to Gaston Bachelard: «since this spring I have been working on the notion of individuality. This theme seems to me deeply reflective, and therefore philosophical». Simondon states that he had turned his interests, initially directed towards polarity in psychology, towards the theme of individuality. Carrozzini notes that this new research theme cannot but refer to Simondon's readings of the drafts of Georges Canguilhem's *La connaissance de la vie* published in 1952 [1, p. 28]. Moreover, Simondon had probably had many of the writings included in that volume on his desk for quite some time, as they had been written between 1946 and 1947 [1,

30]. These essays also include *Le vivant et son milieu*, the essay in which Canguilhem criticises the extreme positions of Piéron's scientific psychology discussed above.

Canguilhem's thought strongly inspires the Simondonian perspective on several levels. This can already be seen in a text drafted in 1937 [53, pp. 77–85] in which Canguilhem scrutinises Descartes' theses by showing what Lacourt calls a *basculement* [54, p. 65]. If in the *Discourse on Method* Descartes had concluded that practice could not attain the perfection of theory, in *The Dioptrics* he seems to prove the exact opposite when he insists on the role played by the technical imperfections of telescopes in discovering how spectacle lenses would be best constructed. Canguilhem notes this inconsistency and states that:

> Science proceeds from technique [...] in that technical embarrassment, failure and failure invite the mind to question the resistance encountered by human art, to conceive of the obstacle as an object independent of human desires and to seek true knowledge [53, p. 92].

The theory, therefore, is developed subsequent to the implementation of techniques designed to demonstrate to human beings the usefulness of error. Simondon's contributions to this field are significant, as evidenced by his seminal work on the history of technique, particularly in the lectures he delivered at the University of Poitiers in1960–61 [5, pp. 27–130] and at the General Psychology and Technology Laboratory he directed at Paris V in 1970 [5, pp. 131–178].

Canguilhem's reflections, which turned towards medicine and biology in the 1940s, resulted in a thesis for his medical degree in 1943 entitled *Essai sur quelques problèmes concernant le normal et le pathologique*, which was later summarised in *Le normale et le pathologique* [55]. In the latter text, Canguilhem expounds on the conception of the *pathological* understood not as a quantitative variant of the *normal*, according to the conception of positivism, but as an «other normality», in which health and healing do not signify the re-establishment of some previous status quo, but rather the quest for a novel equilibrium. According to Canguilhem, the *normal* state does *not describe* anything but prescribes a condition of health. The proposal is for a reunification of physiology and pathology in a *science of health conditions*. From the very pages of 1943, Canguilhem emerges as a scholar *sui generis*, akin to the ancients, in his unique integration of philosophical reflection and medical technique, charting an alternative course away from the prevailing existentialism, phenomenology and spiritualism in academic circles. In the afterword to the first edition of 1966, Foucault stated that Canguilhem was the reference point for all French thinkers of that period:

> Remove Canguilhem and you will no longer understand much about a whole series of discussions between French Marxists, nor will you grasp the specificity of sociologists such as Bourdieu, Castel, Passeron, who leave such a strong mark on the field of sociology; you will miss out on a whole section of the theoretical work of psychoanalysts and Lacanians in particular [55, p. 271–272].

In the 1943 text, Canguilhem vigorously criticises the work of Claude Bernard (an author subsequently explored by Simondon, particularly in the second section of *ILFI*) for his conceptualisation of disease as a distinct state from health, which Bernard claimed would not engender any novel developments within an organism. In contrast, Canguilhem's perspective emphasises that disease or pathology signifies the potential for a transformation within the living organism, resulting in the emergence of novel norms that differ from those prevailing in the preceding state of health. When a living being recovers from an illness, it does not, in fact, return to the state prior to falling ill, but enters into a different and always momentary equilibrium as a new pathology may arise to be faced. The individual who professes illness and seeks medical attention does so in relation to a specific *aspect* [54, p. 37][9] of life to which they were formerly accustomed and now feel distanced. Canguilhem asserts that «to be sick is really to live another life for man, even in the biological sense of the term» [55, p. 49].

During the 1940s, Canguilhem cultivated a profound admiration for Gaston Bachelard, whom he succeeded to as professor of Histoire de la science at the Sorbonne in 1955 [56]. The 1955 is also the year in which Canguilhem had discuss, with Bachelard as supervisor, his thesis in philosophy on the history of the notion of reflex—from which Simondon will take his cue for his course on instinct and his non-mechanistic conception of this concept. Canguilhem states that:

> Today we live in a form of civilisation that has given speed and the automatism of motor reactions a double value: utility and efficiency for the machinist, prestige for the sportsman. The reflex, therefore, is no longer just a scientific fact known to specialists, but is, so to speak, a fact of public utility and public notoriety [56, pp. 162–163].

He dedicated several articles and essays to the epistemologist, from which three main axioms of Bachelard's thought emerge: the theoretical primacy of error [57], the speculative disregard of intuition [23], and the object as a perspective for ideas (Bachelard [93]). Bachelard's historical epistemology profoundly influenced Canguilhem, enabling him to develop a more comprehensive and nuanced understanding of the subject.

The influence of Bachelard and the problem of the formation of scientific concepts can also be seen in the collection entitled *La connaissance de la vie*, published in 1952. From these essays emerges the work in physiological psychology carried out by Canguilhem; in particular, the emphasis on the importance of the transition from what can be described as a Cartesian mechanistic reflexology to the animal psychology of behaviourism [54, p. 61]. Canguilhem links this shift to another innovation brought about by Georges Friedmann: the first study of social ethnography in the Western world made by comparing different points of view (mechanistic, biological, psychological, sociological) (Friedmann [94]). Canguilhem finds in Friedmann an important analysis of how, with the advent of industrialisation, the inventive character of the human being as an individual who «composes his environment according to a way of thinking that he cannot renounce without depriving his life of all meaning» was gradually abandoned [54, p. 62]. It is a theme, that of the misrepresentation of the

[9] Lecourt points out that Canguilhem uses the term allure, translatable as gait, semblance or appearance: «it is movement according to more or less (fast) but also according to rhythm, it is the step, the trot, the gallop». These *allures* can never be totally stable, but stabilise from the breaking of a previous *allure*.

human being in the world of work, which will intensely animate Simondon's work criticising automatism and the labour system.

In the collection of essays published in 1952, Canguilhem brings up an issue that will later inspire Simondon: «one must "understand" matter in life and the science of matter, which is science simply, in the activity of the living being» [56, p. 141]. This statement seems to point the young Simondon in the direction of the primacy of processuality in the search for individuality. Canguilhem wants to say that the study of life and the search for truth, proper to science as to philosophy [58], must participate of the very processes of the living being's activity, beyond mere observation. In this sense Canguilhem seems to recall those same discoveries of quantum mechanics that Simondon relaunches in *ILFI,* which emphasized the impossibility of dividing the knowing subject from the known object. In this sense, Canguilhem suggests that the correlation that exists between what is defined as subject and object comes, in truth, before the very definition of the two poles. It is impossible not to trace back to these early efforts the ontology of relations that Simondon would develop.

2.8 1958: Reconceptualising Reality Between *MEOT* and *ILFI*

On 26 April 1958, at the age of thirty-four, Simondon defended his two doctoral theses in front of two prestigious colleges of professors in order to obtain a State Doctorate. In France, it was customary to prepare a more substantial main thesis and another called a complementary thesis. The panel for the main thesis included Jean Hyppolite, Paul Fraisse and Paul Ricoeur, while the complementary thesis panel comprised Georges Canguilhem and Raymond Aron. In attendance at the proceedings were two other key figures in Simondon's work: Maurice Merleau-Ponty and Martial Guéroult.

On this occasion, the theses presented by Simondon generated several discussions and questions among those present, as Fraisse [28] himself attests, and some dissent, although it was then decided to vote for an honourable mention for the research developed in the theses [1, p. 30; 59].[10] In particular, different discussions developed around the main thesis that Simondon presented, albeit only partially. In truth, Simondon's project was much more extensive, yet he only defends the initial phase of his research. The second part of the research, however, is not discussed and will only be published in 1989, the year of the philosopher's death, when the Aubier publishing house decided to make it a separate book (*IPC*) from the one dedicated to the first part, which was published previously [88]. The two parts of the thesis will only be reunited in 2005, the year of publication of the complete version of *L'individuation à la lumière des notions de forme et d'information* (*ILFI*). The publishing history

[10] Carrozzini reports that these statements by Fraisse could be vitiated by the intellectual tension that was generated between Simondon and Fraisse himself in the years they spent together at the Sorbonne, as reported in the biography of Fraisse compiled by Serge Nicolas.

of Simondon's main doctoral thesis is very complex and contributed to his estrangement from philosophical academia. After its defence in 1958, the main doctoral thesis was not as successful as *MEOT* (published as early as 1958) probably due to the complexity and technicality of some sections. One has to wait until 1964 to see the publication for the Parisian edition Puf in the section edited by Jean Hyppolite of a first part of Simondon's enormous work. It is entitled *L'individu et sa genèse physico-biologique* and brings together the introduction, conclusion, the first two chapters of the first part and the entire second part of what is now the complete work. This was to be the volume that arrived on Gilles Deleuze's desk and of which he published a review in 1966 [85] (pp. 115–118). The third and fourth parts of *ILFI* only came to publication in 1989 for the publishing house Aubier (*IGP*) in which the 1960 essay *Forme, information, potentiels* was also added as an additional paragraph to the Introduction: *L'individuation psychique et collective*. In 1995, Millon publishing house decided to republish *IPC* with the addition of the third chapter of the first part, previously omitted at Simondon's behest, three supplements (*Analyse des critères de l'individuation, Allagmatique, Théorie de l'acte analogique*), and Jacques Garelli's preface. Only ten years later, in 2005, did Millon publish *ILFI*, the complete version of the 1958 thesis, with the various additions made over the years and the unpublished *Histoire de la notion d'individu*, which was originally supposed to be the introductory chapter of the thesis (as stated in the typewritten notes included with the original version of the thesis preserved in the Canguilhem archives of the ENS in Paris).

The history of the complementary thesis is much less troubled. It is thanks to its publication as early as 1958, a few months after the defence, that Simondon's thought will for a long time only be associated with a reflection on technique. After 1958, *Du mode d'existance des objects techniques* (*MEOT*) was republished several times (1969, 1989, 2001 and 2012), always by the Parisian publishing house Aubier, and was quite successful among the ranks of technical philosophers who saw in Simondon's reflection an originality and a novel proposition regarding the question of the relationship between man and machine.

To 1971 dates the first attempt by the critics to connect the two research. Érik [60, pp. 47–62] was, to all intents and purposes, the first to understand the value of Simondon's analysis of the technical object as a way to frame in general problem of individuation, in the context of a larger attempt to rehabilitate the question of becoming. Laloy argues that Simondon sought to «progressively constitute paradigms that allow us to think systems in becoming» [60, p. 48]. And technique is that sphere of operations, generations, transformations and evolutions that can be made to walk alongside nature rather than against it, in order to understand the nature of the living and, specifically, the human. In Laloy, there is an attention to the terms and examples brought by Simondon (the crystallisation, the membrane, the relation, the retinal disparation) in order to get to the bottom of the connection between the two research. He especially notes Simondon's emphasis in both works on operation and relation as primary to any possible structuration, organisation or individuation. Laloy's analysis, however, lacks the part on psychic and collective individuation. This gap is filled in 1989, the year *IPC* was published, by the commemorative text by Denise Van Caneghem, Simondon's student [61]. The latter retraces Simondon's research with particular emphasis on the treatment of the themes of anxiety and memory as an externalisation of our body's function as an object among objects [61, p. 820]. Another point highlighted by Van Caneghem is the type of control and power exerted by the opacity of the technical object in the different stages of structuring culture, as noted by Simondon in *MEOT* [61, p. 827].

In the aftermath of Simondon's untimely death and Van Caneghem's tribute, a special issue of the journal *Cahiers Philosophiques* appeared in 1990 in which some of his students participated (VV. AA. [95]).[11]

From the early 1990s onwards, interest in Simondon's research began to grow thanks to the conference organised in his honour at the *Bibliothèque du Collège International de philosophie* in April 1992. The proceedings of this conference, published in 1994 [62], include two of Simondon's writings (one unpublished and an article from 1959), and among those commenting on his research appears for the first time Bernard Stiegler [62]. The same year saw the publication of the first volume of *La technique et le temps* in which Stiegler already shows a profound knowledge of Simondon's thought in its entirety. He is perhaps the first to have found the key to holding *MEOT* and *ILFI* together in temporality. This is an aspect that often appears in Simondon's treatment, but never thematised as Stiegler did with the support of Heidegger's and Derrida's theses. Stiegler would later return to Simondon several times. First in 1996 with the second volume of *La technique e le temps* (1996) [91] and then in 1998 with: *Temps et individuations technique, psychique et collective dans l'œuvre de Simondon* [92, pp. 241–256]. In the *Introduction* Gilles Châtelet states that Simondon had been far-sighted when he argued in the introduction to his main thesis that:

> In order to overcome the traditional dualisms: individual-community, form-matter, it should not give excessive privilege to the accomplished individual but should resolutely tackle the problem of individuation [62].

Through this insight, Simondon had paved the way for a new conception of reality at the centre of which was the need to reconceptualise individuality. A project that the philosopher shared with the world of science. Dating back to the 1950s, in fact, was the discussion between Pierre Auger and Erwin Schrödinger in which the latter stated that one of the questions to be asked was that of individuality, since its identification with the person could only be a mere illusion [64].

The new path proposed by Simondon implied the abandonment of any kind of dualism in favour of a primal relationality, but also a profound attention towards the scientific, technical and technological world through whose discoveries it would be

[11] Within this issue you can find a text by Simondon entitled *L'effet de halo en matière technique: vers une stratégie de la publicité* originally published in 1960 in the journal *Cahiers de l'Institut de science économique appliquée.* The following essays are also included: John Hart, *Les catégories remarquables de la médiation technique*; Jean-Yves Goffi, *Gilbert Simondon et Jean Brun sur la technique*; Jean-Claude Beaune, *La philosophie des techniques de Gilbert Simondon. Perspectives sur l'individu et l'invention*; André Pichot, *L'individuation biologique*; Bernard Balan, *La philosophie de la nature de Gilbert Simondon*; Jean-Yves Chateau, *Technologie et ontologie dans la philosophie de Gilbert Simondon.*

possible to trace the concepts necessary for the foundation of a new ontology. As already stated, a fundamental role must have been played by Simondon's knowledge and in-depth study of Wienerian cybernetics and its developments in a French environment. Indeed, the years immediately preceding his doctoral defence are the years in which the philosopher dedicated his writings to cybernetics [33], and the organisation of the Royamount conference on the concept of information [65] with the support of Merleau-Ponty, which dates back to the decade 1950–1960.

Starting from the centrality of the concept of control in cybernetics, Simondon thought it necessary to rethink the latter in order to study the relationship between living beings and machines. Indeed, the history of the genesis and evolution of technical objects, which Simondon reports in the first part of *MEOT*, tells of a complex relationship with the notion of control. Every technical object is generated for a function but contains within it several possibilities of use. Therefore, in order to make the technical object do what you want it to do, you must learn to use it in a certain way, limiting its possibilities and directing them towards the one function you want it to perform. The manufacturer and user of the technical object exercise control over it by imposing on it a choice between the various possible functions it could perform. This possibility of exercising control over the function to be performed by the technical object leads to a sense of omnipotence in the human species that can lead to completely forgetting the presence of *open virtuality* (the various possible functions) in technical objects. In reality, even with the restriction of the possibilities of the technical object by the human user, this virtuality persists, as a zone of shadow within the object that makes it, over time, the symbol of danger, insofar as it is impossible to control since it cannot be definitively identified (irreducible plurality). According to Simondon, the drive to control as the possibility of maintaining complete rationality and transparency goes back to Greek philosophy, in particular the idea that instincts, of which organic bodies are made, are the part that man must dominate through reason in order to avoid remaining in mere animality [49]. The reference here is to the idea of the *lesser animal* already present in Democritus and in Plato's version of the Prometheus myth. Epimetheus distributes capacities to all animal species and when he arrives at man, he realises that he has no capacities left to give him. Prometheus then intervenes and, to make up for his brother's mistake, steals fire from the gods and gives it to man (along with the ability to understand its mechanism). Fire represents the technique that until then had been the prerogative of the divine. When Prometheus hands it over to men, he gives them the ability to survive because they lack other abilities. From this idea of man as a lesser animal Plato develops the thesis of the soul as an instance of control present in man to keep him in contact with both the divine and the earthly [47, Ch. 1.5]. Simondon takes up the theme of control on several occasions and links it to the re-evaluation of instincts. In particular, the first chapter of the first part of *ILFI*, dedicated to the relationship between form and matter, highlights the instance of control behind the Platonic and Aristotelian theses in which form brings order, through a structure, to the irrationality of matter (*ILFI*, pp. 39–66). The same theme is taken up in the context of a history of the notion of the individual in the Complement to the complete version of *ILFI* (pp. 357–522). Furthermore, Simondon dedicates a course in general psychology in 1964 to instinct, as an apparent advocate of the need for the rational aspect to prevail over the irrational in order to aspire to wisdom and a happy life. In this course, Simondon argues that instincts are neither sets of segmented irrational elements (mechanism thesis) nor the primitive form of rational/intelligent behaviour (finalism thesis). Instincts, for Simondon, are connected to both rationality and irrationality because they are like relays that make motor and energy conditions interact with perception [66, pp. 75–103, 67, pp. 247–374]. The instinctual/pulsional part of the human being nevertheless remains present. Through technical objects, man attempts to gather rationality and freeze it. The technical object thus becomes, in a way, man's *rational testament*. But if the technical object is created by a man that has two sides, logical-rational and emotional-rational it is not possible to think that only one of the two parts can be transferred into it. Despite this, man actually thinks he can construct fully automatic technical objects, i.e. entirely logical-rational objects that are never obstructed by alternative possibilities. If, however, a technical object was constructed without alternative possibilities such, such object would not need external control because it would already have the autonomy necessary for its own control. This type of completely rational technical object would thus entail (and in fact has entailed and entails) the image of a disturbing and threatening *automaton* taking control of everything non-rational.

Thus, a short circuit emerges when man tries in every way to make his rationality prevail in order to avoid falling prey to his emotions, in fear of being controlled by something external. The technical object is constructed by man as the centre of attraction of pure rationality. But precisely when it is thought of as pure rationality, it is not really controllable by any entity. And so, the instance of control that has guided the history of mankind turns against man, who becomes prey to the fear of being controlled by those entities he had constructed to preserve his rationality.

As proof of the central role of this intuition, Simondon will dedicate a semester of his general psychology courses at the Sorbonne in 1964 [66][12] to instinct, a key term in the reversal of the idea of control. In it, the philosopher stated that only a psychological study of drives could clarify the characteristics of instinct without reducing it to reflex, tropism, or simple spontaneous movement. In these lectures, Simondon will recall that most behaviours are, in truth, accumulations of instinctive processes that, through learning, have stabilised. These processes are activated whenever homeostatic functions reach their limit and risk being overwhelmed. For Simondon, the study of instincts should be preceded by an investigation into perception, in order to clarify the mediating role (*relè*) that instincts play. Research, therefore, must be conducted at the psychological level to identify the roots of the misunderstanding

[12] Extremely interesting from a historical and reconstructive point of view of the cultural context in which Simondon moved is the list of bibliographical texts given at the beginning of the course transcript, the following volumes: H. Bergson, *L'evolution créatrice*; M. Merleau-Ponty, *La structure du comportament* [86] e *Phénoménologie de la perception* [87]; K. Goldstein, *La structure de l'organisme*; N. Wiener, *Cybernetics*, H. Piéron, *Psychologie zoologique*; Rabaud, *L'instinct et le comportament animal*.

concerning the function of control. Only by revising this perspective and reintegrating instincts within the framework of individuality can we hope to reorganize our entire understanding of reality.

It is for this reason that we have chosen to start from the general analysis of the work on the genesis and evolution of technical objects in *MEOT* and then move on to the third part of *ILFI*, the one dedicated to psychic individuation, and, finally, to analyse the 1960 article on *Form, Information and Potentials* as the point of contact between the two works. It is a choice dictated by the interception of a need, primarily Simondon's, to hold together the visible and the concealed reality, the non-given, whose first representation, in human reflection, is precisely that of the psyche. Only by starting from the analysis of this first representation of invisible reality is it possible to arrive at the relationship that binds these two levels of reality and annul the rational/logical objectivising limit that denies effective freedom in the name of control/domination.

2.8.1 Du Mode D'existence Des Objects Techniques

The theme running through the work of Simondon's supplementary dissertation takes its cue from a need that has been present in the reflection on technology since the work of Karl Marx: trying to understand the transformations of society since the advent of technological progress following the industrial revolution. Authors such as [68], Lewis [69] and, with greater reference to more recent technological innovations, philosophers belonging to the Frankfurt School such as [70, 71]) or students of Heidegger such as Günther [72] (or the later Herbert Marcuse and Hans Jonas) had in various ways argued that the relationship between human beings and technology would undergo a radical change with the advent of industrial technologies, provoking a network of feelings that Simondon would define as *disruptive* and, ultimately, opening the way to alienation and anxiety. Simondon takes the term *disruptive* once again from the scientific world, in particular from electronics, where the adjective disruptive refers to an electrical discharge that is established in an electrical insulator when the electric field in it exceeds the dielectric rigidity and that in a solid medium can lead to the actual rupture of it. In a note commenting on the third chapter of the first part of *ILFI*, dedicated to form and substance, Carrozzini refers to the phenomenon of disruption in the case of the ionisation of a gaseous dielectric [1, p. 70]. At the end of the 1960 essay *Form, Information, potentials*, regarding the *disruptive* character of an operation within a domain Simondon states that it is the opposite of a transductive operation since it does not generate structure but only destruction (*ILFI*, p. 533). It is possible to trace a common matrix to these positions in the idea (clarified in detail by contemporary philosophers of technology such as Bernard Stiegler) that the dominance of machines would trigger a process of displacement of human beings' internal temporal structures due to the acceleration of production times. Following the logic of control that we have previously expounded, far from feeling controlled by the machines, the human being would henceforth ceaselessly seek to heal the gap between his organic times and those of the machine.[13]

The asynchrony between these two times would, however, be impossible to heal because the rhythm of technology has always been faster than the organic one. According to these authors, due to technological innovation, man would feel the need to chase technical time to such an extent that he would become pathologically trapped in a sense of impotence from which he would no longer be able to emerge. It is starting from this idea that these authors considered industrial society to be pathological, in the sense of Marcuse's "one-dimensional society", i.e. inexorably entangled in apparent technical improvements that, in reality, would only lead to a backwardness and annulment of the human. In this view, every technological innovation/evolution would be one more step in accelerating a paradoxical *involution* of the human.

Simondon's first trait of originality within this panorama consists in his taking a different perspective early on. He develops an interest in the technical object whose aim is neither to exalt nor to destroy it. In fact, *MEOT*'s aim is to «show that the culture ignores a human reality within technical reality» (p. 15) and that the technical object is not a stranger seeking to gain power and control over man, but an otherness that mediates between the human being and nature. It is on the basis of this innovative premise that Simondon considers it necessary to examine the genesis and evolution of technical objects in order to dispel the myth of the robot that would like to take the place of man, and to demonstrate how the mismatch between technical and organic times is not actually an internal distortion of the living, but rather a fundamental dynamic of technical-biological evolution.

In contrast to authors such as Anders or Marcuse, Simondon argues that the cause of man's alienation does not lie in his being the victim of rational control by machines, but in his lack of knowledge about them (*MEOT*, p. 15). Through an in-depth analysis of the processes of genesis of technical objects and their evolution, Simondon comes to the conclusion that every technical object, preserving the *realitè humaine* of its maker, cannot be, as the latter is not, completely automatic. Within every technical object there is always structurally present a non-automatic or *de-automatic* component that allows the human being who uses it to see it as a support in his relationship with the world. If the technical object were not simultaneously automatic and de-automatic, it could not in fact communicate with the human being, making itself a mediator with something else, that is to say, a means of disengaging the automaticity of a pre-existing human–environment relationship in favour of a new chain of connections.

In this sense, Simondon states that the condition for a technical phylogenetic line to be realised is that there is a primitive stage that tends towards a concrete stage of the technical object, as the latter only exists as a unit in the making in which a series of convergences come together. The primitive stage is represented, according to Simondon, by the technical object constructed by the craftsman insofar as there is

[13] In Anders it is called "Promethean unevenness".

still no internal coherence in it. This absence is determined by the fact that its form is dictated by the will of the individual human being, it is «customized» (*MEOT*, p. 30). By contrast, the concrete stage of the technical object can be found in the industrial object for which it is not external needs that determine its conformation, but the opposite. The transition from the primitive or abstract stage to the concrete stage enables the evolutionary movement of the technical object. But for this to occur, the abstract object must be imperfect. In fact, the evolutionary movement is triggered when an abstract technical object does not meet properly technical requirements, such as not having to be self-destructive and having to remain in stable function for as long as possible.

So the evolution of technical objects can only occur if there is a deviation between the structure of the abstract object and the technical requirements, i.e. if there is a problem to be solved consisting of the need to jump over a previous automatism (a stable connection between operational procedures) in order to realise a new one. The resolution, argues Simondon, «can occur only as a leap» (*MEOT*, p. 32) that allows an internal reorganisation of the system of functions of the abstract object, enabling the realisation of a concrete object. It can be said, therefore, that technical evolution follows two lines: an internal redistribution of functions that allows for greater compatibility (*differentiation*) and an internal reorganisation that allows each operation to produce effects that will not harm the overall operation (*concretisation*).

The technical evolution described by Simondon, therefore, proceeds by stages of synergetic association, but each discovery of functional synergies «happens abruptly» (*MEOT*, p. 40). In this sense, the genesis of a technical object is not just a path from the abstract to the concrete, but an evolution that takes place by discontinuous essential refinements that modify the internal system through leaps and bounds. The minor, continuous perfections have the role of overloading the essential structures of the abstract technical object, ultimately leading to a discontinuous leap of concretisation. To explain this coexistence between continuity and discontinuity, Simondon resorts to the term *mutation* and states that «the true stages of the technical objects improvement occur through mutations» (*MEOT*, p. 43). Therefore, for Simondon, technical evolution moves by sudden and discontinuous mutations that are oriented through the action of those minor refinements that, one might say, filter the possibilities of major refinements.

Starting from a mismatch (with the abstract technical object), the concrete technical object evolves by adapting to technical and economic requirements. However, there is the possibility of a suspension of the evolutionary path for certain structures that fail to materialise immediately into a synergetic system of functions. In this case, says Simondon, one does not lose the refinements one has found, but these remain suspended pending future readaptation in another sphere. We could therefore say that in Simondon, the adaptations that occur with a minor refinement or a major one are never definitive; the technical object always remains a unit in the making, ready to de-adapt and re-adapt every time the internal synergetic system breaks down.

Another essential passage in Simondon's description concerns what happens to the technical object once it has materialised: «technical concretization makes the primitively artificial object increasingly similar to a natural object» as it frees itself from the necessity of the «external regulating environment» (*MEOT*, p. 50) and incorporates it through its association with the other technical or natural objects that surround it. The incorporation of the environment by the technical object allows it to evolve by adapting itself not so much to a certain type of environment,[14] but to the function capable of relating two different environments in the midst of which the object finds itself:

> Adaptation-concretization is a process that conditions the birth of a milieu rather than being conditioned by an already given milieu; it is conditioned by a milieu that only exists virtually before invention (*MEOT*, p. 58).

In order for an evolution of technical objects free from over-specialisation (*hypertelia*) to take place, the creation of a third element in the relationship is needed, namely the techno-geographical environment. This is an environment that lies between man and nature and that the human being can create because he has the inventive capacity (*MEOT*, p. 59) to make a new structure appear from an already given reality. In the invention of the associated environment (techno-geographical), which will lead to that of the technical object, man does not carry out a synthesis of the structures present in the given reality, as *Gestaltpsychologie* had hypothesised, but moves by leaps and bounds. The human being, as an individual always in relation to his associated environment (*ILFI*, second part), disassembles the mechanical-automatic unity of the starting environment by assembling the technical elements that do not have a coherence between them and foreseeing or imagining the future coherence in relation to the techno-geographical environment associated with the future technical object. This is a process that can occur in the absence of man, but only on rare occasions and only by chance: for example, the first *aeolite*[15] found by man was shaped to be used as a stone tool. Human beings, on the other hand, are able to carry out this process, when necessary, thanks to the function of imagination, linked to that of invention of ever new structures.

Simondon emphasises how important it is to analyse the inventive imagination in humans from a psychological point of view [40] and how little attention has been paid to the role of the background, especially by *Gestaltpsychologie.* Until then, psychologists' gaze had always been pointed towards the pattern or form generated by perception and imagination, to which the main energetic force had been attributed. Simondon argues, on the other hand, that it is precisely the background that gives energetic force to forms as a reservoir of the virtuality of possible forms (*MEOT*, p. 61). Forms are passive as long as they do not enter into relation with the background and become active when they allow the background's system of virtuality to pass into the actuality of the pattern/form that is identified.

---

[14] Simondon calls the phenomenon of adaptation by specialisation to a specific environment *hypertelia*.

[15] In the second chapter of *Man the tool-maker* of 1949 Kenneth Oakley describes aeolites as: «the earliest recognizable implements and represented the dawn of tool-making [...] eoliths can be matched exactly by stones chipped by natural agencies». In any case, it is not possible to distinguish between aeoliths chipped by man and those shaped by environmental conditions. See Oakley 1961 (5th edition), pp. 5–7 ff.

Simondon emphasises the link between the human being and the technical object through the parallelism between the functioning of the interaction between background and forms in the inventive imagination and the «relation of each of the technical objects in structures which are constituted with the dynamisms of the associated milieu» (*MEOT*, p. 61). Between the structures of the technical object and the associated techno-geographical environment, an asymmetrical causal recurrence is repeatedly nurtured in which the environment plays the role of the background as the seat of self-regulation, while the structures move without following the causality of the background, «each in its own sense» (*MEOT*, p. 61). The mismatch between the associated environment (background) and structures (forms), both in the psychic life of the living being and in the technical sphere, leads to alienation insofar as the former is no longer able to regulate the dynamism of the latter.

Thus, human beings can invent objects because their psychic life functions according to this relational process between a background (thought) and forms (life or discontinuous representations):

> We can create technical beings because we have within us a play of relations and a matter-form relation that is highly analogous to the one we constitute in the technical object. The relation between thought and life is analogous to the relation between the structured technical object and the natural milieu (*MEOT*, p. 62).

In the absence of this interplay between life and thought, the constructed object is merely a tool since it responds to only one of the two systems and lacks, instead, the associated techno-geographical environment that would allow it to be a concretised technical object open to evolution. Therefore, Simondon refers to the need to analyse psychic life in greater depth and, in particular, the dynamics of perception and imagination as functions that cannot be mere inductive or deductive processes that lead back to pre-existing structures. Thus, it is essential to connect this need expressed in *MEOT* with the third part of *ILFI* devoted precisely to reconstructing the processes of psychic individuation.

2.8.2 Psychic Individuation

The third part of *ILFI*, first published in France in 1989 (*IPC*), despite not being among the papers presented in 1958 during the thesis defence, was drafted by Simondon in the same years as the first two parts. The pages on psychic individuation were certainly written around the time Simondon was lecturing on modern psychology in Poitiers.

Simondon's first objective is to analyse the problem of the segregation of perceptual units and to understand how the subject «perceives objects as having an already given and consistent individuality» (*ILFI*, p. 257). Simondon's choice to start from perception will be justified in the courses he held during the 1960s at the Sorbonne, in which it can be seen how, in his opinion, there is no real continuity between the faculties of living organisms and how, instead, one proceeds by accumulation of a potential energy that is amplified by one of the faculties in an attempt to solve the problem posed in the previous phase. In particular, Renaud Barbaras emphasises the fact that perception is, for Simondon, a central node for psychology (and consequently, for the entire ontological infrastructure) since it represents a privileged mode, defined as *living*, of the relationship between man and his world [73]. For Simondon, perception is not contemplation and should not be studied from the point of view of its contribution to knowledge, but an activity to be investigated under the biological and informational gaze. In this perspective, the object of perception loses its objectification because it is never a stable individual, but is a process that enters into relation with another process. Picking up on Pièron's indications, Simondon argues that one cannot study the perceptual process without taking into account that, as a privileged mode of relation between man and his world, it intervenes in further processes whose affective/emotional scope orients perception itself.

In this text and later in the course he gave in '68 at the *ENS*, Simondon does not fail to emphasise that the primary characteristic of the living organism is operativity and that perception is a manifestation of the latter. It is a manifestation in which operativity translates into a modulating or amplifying function of stimuli. The idea of perception as a paradigm of operativity is conceived by Simondon from the observation of the amplifying and modulating process within techniques and from the lesson of his teacher Georges Canguilhem. The latter, in *La connaissance de la vie* (1965), affirms that *the vital is an irreducible originality* and that knowledge, reflection, arrives later than the other characteristics of the human being as a living being, since it arrives as a detachment of the vital individual from the world, allowing solutions to be found to the problems that arise within this relationship.

Although perception remains for Simondon the central knot to be unravelled even in relation to the interpretations given by the theories preceding or contemporary to him, in the thesis this faculty does not occupy the role of the original relationship between organism and environment since prior to all perception there is the motricity that allows the organism to be within a spatiotemporal system.

The primacy of perception in Simondon, therefore, is not comparable to that assigned to it by French phenomenology: it is primacy as paradigm, for understanding analogically the amplifying functioning of other endo-organic activities.

Picking up on the criticism of associationism already present in '56, Simondon states that the associationists' answer to the question of the segregation of perceptual units lies in the role of habit, which would allow the units of which the perceived object is composed to be communicated to perception. This thesis, however, conceals, according to Simondon, a profound innatism insofar as the mere collection of the units of which an object is composed would render it a «mere accumulation» without any internal coherence: association by continuity, therefore, presupposes an innate coherence in the object (*ILFI*, p. 257).

The same innatism would not, however, be concealed in *Gestaltpsychologie*, which explains the segregation of perceptual units through recourse to the laws (innate in fact) of *prägnanz* and good form, which would allow the subject to grasp the unity of the object immediately. For Simondon, as in associationism, the problem is posed in the wrong way insofar as it is assumed that there is something pre-formed. Gestalt psychology is a constant reference for Simondon both for the concept of the field that is illuminating in his eyes and for the negative aspects concerning the excessive emphasis on good form as a condition of stable equilibrium. It is a psychological theory whose theoretical foundations can be traced back to Franz Brentano and his 'psychology without a soul' in which an attempt is made to abandon the study of the psyche in order to give psychology the status of an objective science. With Brentano, therefore, the focus shifts from the psyche to perception. From what is called the «Brentano school» emerged the core of *Gestaltpsychologie* theorists, including Carl Stumpf—who went on to found the Berlin school in which the second generation of Gestaltists was formed—and Christian von Ehrenfels [74, pp. 183–241]. It was von Ehrenfels, taking it from Ernst Mach, who introduced the term *Gestalt*, which then gave the theory its name. *Gestalt* properly means 'form' and comes from the verb gestalten which indicates the action of giving form to something. The term would later be taken up by the Berlin School and Stumpf's followers to give birth to *Gestaltpsychologie*, a psychological theory that sought to oppose associationism and affirm that the whole is more than the sum of the parts. This translates into a greater focus on the act of perceiving as a moment in which not only sensations are added up, but something more is added. *Gestalt* is, therefore, this more, that is, the ability to organise elementary sensations into figures emerging from a background. In this way, we obtain a figure with detailed contours, which emerges sharply from an undifferentiated background, which in some cases appears imperceptible. Recalling Brentano and the need to make psychology objective, *Gestaltpsychologie* theorists propose certain laws that perception would follow to obtain this figure emerging from the background: the law of poignancy, the law of similarity and the law of good form [75, 76, pp. 235–248].

It is here that Simondon again recalls the problem that animates the entire research: «a rigorous study of individuation, i.e. this critical moment when unity and coherence appear» (*ILFI*, p. 258) has been lacking in both associationism and *Gestaltpsychologie*. It is from this that we must start in order to understand, according to Simondon, the first step in the process of psychic individuation, namely the segregation of perceptual units.

This process takes place in what Kurt Lewin calls the psychological field,[16] whose composition would consist of the three elements subject, world, and the relationship between them. Perception, as a process of individuation, aims to resolve the tension that develops between these elements. Lewin allows a way out of the impasse of behaviourism by providing the *Gestaltpsychologie* theory with the concept of *field*. In the latter theory, however, there remains the idea that tension is absent within the process of perception because the laws of good form, which are its presupposition, do not require a problem to be solved in order for a form to be grasped. In fact, Simondon argues, for there to be psychic individuation in terms of perception there

[16] Lewin turns out to be one of Simondon's major references as the bearer, within psychology, of the notion of field elaborated within physics. This notion will be of fundamental importance as it will make it possible to overcome behaviourism and think of the relationship between the individual and the environment as a totality in which the individual parts cannot be studied separately [77].

must be a certain «degree of metastability» (*ILFI*, p. 258) that cannot be found in the perfect geometric forms identified by *Gestaltpsychologie* as the most *Prägnanz* forms. This thesis is also developed from an experiment carried out by zoologist Adolf Portmann on a child's perception of a lion (1952). Although the process of perception only occurs once, the child is left with the figure of the lion not because it can be framed within the structural schemes of good geometric shapes, but because he perceives the different parts of its body due to a situation of tension (positive or negative) in which he finds himself and which leads him to find a solution. The latter is given precisely by the lion's body scheme that reorients the perceptive field of the child's body scheme.

This experiment leads Simondon to claim that only a state of metastability can allow a process of psychic individuation, i.e. a perception of unity. The concept of *metastability* had already been presented by Simondon in earlier parts of his research and is taken from physics, in particular the crystallisation process. It is a state of tension in which the main actors are the relationship and the potentially different responses that can emerge from the relationship. The living individual, unlike the physical individual, does not free itself from metastability and, on the contrary, continues to perpetuate it in a continuous attempt to solve the problems posed by successive spatial and temporal structuring that specialise the individual irreversibly.

In order to analyse and understand how a psychic individuation can occur, Simondon resorts to a concept already used in the second part, the one dedicated to living beings, taken up again in this case by contemporary physics: the *quantum* (*ILFI*, p. 264).[17] When a subject perceives an object, what he individually grasps is none other than the «reality of certain thresholds of intensity and quality maintained by objects» (*ILFI*, p. 264), i.e. the instant of coherence of a bundle of relations of which the object itself is composed. Now, the subject in question cannot be considered as detached from reality, but is in a concrete situation, immersed in its emotionality (made up of tendencies, instincts and passions). For this reason, Simondon examines both the theory of form and the classical theory of information and expresses the need to find a new theory that does not limit itself to considering only the qualitative aspect of form or the quantitative aspect of information. For a subject immersed in emotional reality, perceiving does not mean grasping the simplest form or the most complex form as the most meaningful, but taking these structurations in their relation to the subject itself in order to be able to orientate itself in the world. The object is a mediator between subject and world, it allows the coupling between them and their polarisation so that the situation makes sense. In the process of perception, therefore, the subject increases the intensity of information, that is, the mediation

[17] The reference is to the thesis presented by Max Planck in 1900 according to which the emission and absorption of radiation by a black body manifested behaviour that was impossible to explain by following the laws of classical mechanics. Planck realised that the black body absorbed and emitted radiation in a non-continuous manner and, therefore, that radiation could not be completely continuous in nature, but neither could it be completely discontinuous. Radiation therefore has a quantum nature, i.e. a discontinuous quantity that emerges from the product of the radiation frequency and Planck's constant ($6.626 \times 10–27$).

between quality and quantity of information that, rather than grasping totality, it organises by introducing a new structure. Perception, therefore, is not a process of collecting forms, but a process of inventing forms from a «recurring relationship between the subject and the world» in which all the terms of the relationship are part of the same system (*ILFI*, pp. 270–271).

It is through this coupling between subject and world that it is possible to solve the problem posed in the relationship between the two, through one or more solutions that stabilise perception, at least for a few moments. Stabilising a perception, however, does not mean making it *Prägnanz*. It is possible to have perceptual stability without it being, in fact, *prägnanz* for that specific subject. *Prägnanz*, in fact, depends not so much on the coupling (and, therefore, on good structure/form), but on the degree of greater or lesser intensity of the initial incompatibility and the solution provided.

For Simondon, it is necessary to pose the problem of the relationship between consciousness and the living individual in order to find an explanation for the ability to segregate units in the world that become objects of perception. In order to analyse this relationship, Simondon proposes to go beyond the gaze of *Gestaltpsychologie*, which only focused on the perceptual relationship, in order to make room for the other aspects of the relationship that concern affectivity. In *ILFI* Simondon draws a distinction between affectivity and emotionality that is often not entirely clear. It is a nuance that, however, exalts affectivity at the expense of emotionality because affectivity 'precedes and follows emotion' and consists in that which allows the individual to connect with both the pre-individual and the possibility of trans-individualisation in the collective (*ILFI*, p. 280) Affectivity, Simondon argues, has a «problematic nature» (*ILFI*, p. 287) and this in the Simondonian setting implies that it is a sphere that can amplify individuation. Thus, affectivity is the problematic sphere through which the individual orientates himself in the environment. Emotionality, on the other hand, is a dimension that opens up once the collective has already trans-individualised and is intended to coordinate and organise «affective universes» (*ILFI*, p. 288). It is important to emphasise, however, that emotion is not the synthesis of affects because once again Simondon does not follow a dialectical logic. Every dislocation and passage to a different level of reality (phase) involves a transduction and, therefore, entry into a third reality different from those that determined its possibility. Indeed, Simondon states that a «more-than-being, a new individuation» is necessary for affects to be co-ordinated into emotions (*ILFI*, pp. 280–291).

Starting from this widening of perspective, Simondon argues that perception, i.e. the segregation of unities, can only take place if the process of individuation continues within the living individual by folding the organism back onto itself and prolonging the same process outside (*ILFI*, p. 273). What emerges in this passage is the inadequacy of the critique brought forward by the second phase of the development of psychology, presented during 1956, with respect to the linking of psychologism to causality. In both Bergson's thought and that of *Gestaltpsychologie*, in fact, consciousness remains an indissoluble and continuous unity (as opposed to the atomist idea of consciousness as an indefinite plurality), lacking the need, noted by Simondon, to hold unity and plurality together. One can only understand the role of consciousness in the process of individuation if one considers that it is quantum, i.e. made up of a continuous, unitary layer and, at the same time, of a plurality of discontinuous jumps in a regime that Simondon calls *transductive*. Affectivity represents the «transductive form par excellence» of consciousness because it allows the relationship that the individual has with himself and that between the individual and the world to be linked.

Psychoanalysis, too, would fail to account for the quantum-transductive regime of consciousness because, according to Simondon, it would identify consciousness as a unity of the psyche in terms of a relationship between the conscious and the unconscious in which the latter would be a kind of cast of the former and, therefore, detectable, controllable and analysable in the same way. Only by considering the unconscious as a layer that influences the individual's reality without being able to be analysed and understood, and the conscious as the manifest layer of summary elements of the unconscious that, by themselves, are incapable of invention, is it possible, according to Simondon, to trace a process that holds these two dimensions together.

Simondon locates this process at the boundary between the conscious and the unconscious in the layer of the subconscious, which constitutes the centre of individuality, being formed by affectivity and emotionality. In this layer one proceeds by leaps, by thresholds that allow one to connect the continuous dimension of the unconscious and the discontinuous dimension of the conscious. Simondon recognises in Jung's psychoanalytic theory a first step in this direction when, in his analysis of the unconscious, the psychiatrist refers to affective/emotional characters that form what he calls the *collective psyche*, i.e. the set of archetypes and myths that underlie the possibility of collective groupings. This means that already for Jung for there to be community it is necessary for there to be an emotional/affective sharing at the basis which, in Simondon, is translated into the mediation between representation and action.

The Simondonian thesis that the centre of individuality is a sphere of affectivity characterised by quantum being also finds support in scientific research concerning the genesis and evolution of organisms and species. As the author acknowledges, the latter assert that all living beings are capable of feeling (affectivity), and that the difference between species lies in the complexity of their nervous systems. However, the basic structure of these systems has a common origin in the midbrain. As a matter of fact, striking this area of the nervous system does, in fact, result in much more serious damage to the personality than if one were to strike areas concerning representation or action. This shows, according to Simondon, that the level to be investigated in order to understand the functioning of the psyche in its relationship with the living individual is precisely the affective one.

Remaining in Simondonian terms, for whom *being in relationship* means communicating, i.e. connecting two levels through the search for a common symbology, the study of the affective layer of subconsciousness entails a new theory of communication. There can be inter-individual communication, according to Simondon, if different individuals can connect through a common symbology. Simondon traces this symbology to the affective symbology common to all living beings; thus, it is possible for there to be communication between a human and an animal of another species that «participate» in what the Greeks called a yoke community (*ILFI*, p. 275).

As was also repeated in the general psychology courses of the 1960s, for there to be perception there must be an affective drive that, after motorisation, is the first type of relationship the organism has with the world. This primary relationship is the level of the real in which the organism fits in as an individual-environment and from which it draws its lifeblood. In the absence of the affective relationship, there can be no individuality. It is through this relationship that the individual can communicate with other individuals and structure what Simondon calls «spiritual life» (*ILFI*, p. 277).

Simondon makes it clear several times, as early as the Introduction to *ILFI*, that any kind of dualism must be excluded from the theory of individuation; therefore, the concept of spirituality he expresses is completely divorced from the idea of spiritual life as opposed to biological/body life. Spirituality, for Simondon, «is not in another life, nor is it the same life» (*ILFI*, p. 278) since it contains within it the unitary and non-unitary being of the individual at the same time. It has been said that the latter emerges, thanks to affectivity, as an individual-environment and, insofar as *it is always in relation*, contains an energetic thrust that requires further individualisation/configuration in order to regain partial harmony. It is precisely because of this need for the individual-environment that the collective or transindividual is configured as the level of reality in which individuated beings with their associated environments find the channels to structure the residual affective/emotional energetic potentials of the relationship, impossible to break, with the preindividual. Before continuing with the analysis, it is important to define the concepts of pre-individual and trans-individual used by Simondon.

The concept of the preindividual in Simondon designates the necessary (and prior) condition for individuation. The preindividual is thus situated before any process of ontogenesis, but as can be seen in *ILFI* this «before» is only a term that serves to explain the process since the temporal dimension also springs from the preindividual: «the preindividual is the source of chronological and topological dimensionality» (*ILFI*, p. 160). This dimension is conceived by Simondon from scientific discoveries such as thermodynamics, the famous wave-corpuscle duality and quantum physics. In the Introduction to *ILFI* he states that these discoveries have made it possible to understand that beyond the individual level, which is what our human logic manages to grasp, there is a deeper level, the preindividual level, from which all processes of individuation, both physical and vital, arise. The critical literature of Simondon's work has rarely embarked on an in-depth analysis of the concept of the pre-individual: it is a complex dimension to define because, as we have seen, it is prior (in the sense of human-temporal logic) to any definition and structuring. Simondon also calls it absolute being to distinguish it from the relative being that enters into each individual, at other times nature in the sense of the apeiron of the pre-Socratics. The only interpretation of this concept that differs from the words used by Simondon is that of Paolo Virno who edited the Italian translation of *IPC* [73]. In the afterword entitled *Moltitudine e principio di individuazione*, in an attempt to identify the preindividual,

Virno first defines it as «the sensorial perception, motility, the biological background of the species» and then as «the natural historical language of one's own community of belonging», somehow betraying that pre-physical and pre-vital presupposition proper to Simondon's definition.

Instead, the concept of the transindividual appears for the first time in *ILFI* in the second part, the one devoted to the vital phase of individuation. In this sense, unlike most of Simondon's commentators who see the transindividual as the union of inner or psychic and collective individuation, it is possible to state that the transindividual is not an operation that occurs only between the psychic and collective phases, but that it is an integral part of every phase of individuation. The transindividual is a different dimension from the interindividual or intersubjectivity because it does not relate two individuals, but is the condition of «three-way» communication between the pre-individual and the individuated. It is interesting to emphasise what is said by Carrozzini in his commentary on *ILFI*: «it is therefore possible to grasp, in the course of the genetic vicissitude of the transindividual, moments of development that first configure it as nascent, and only later as propagating, i.e. prone to the constitution of new individualisations» [35, p. 95]. It must be said that Simondon does not facilitate interpretation by never giving a definition of the trans-individual. This has also allowed commentators to associate the concept with apparently similar ones, as Etienne Balibar has done by relating transindivuality to the constitution of modes in Spinoza [78].

Returning to the analysis, in the movement of individualisation/configuration, spirituality is the foundation of the three-way relationship between pre-individual, individuated and collective, and affectivity, which then translates into emotionality, is the engine of transformation that enables the individual-environment not to remain isolated and flow into the collective:

> affectivity is what leads the charge of pre-individual nature to become the support of collective individuation; it is mediation between that which is pre-individual and that which is individual (*ILFI*, p. 279).

Affectivity regulates, like a relay, the relationship between the pre-individual and the individual, but in order to reach the third of the relationship (the collective) it is necessary for emotion to intervene. In the individual *in relation* to the pre-individual, affectivity and perception are disparate and only through the intervention of emotion is it possible to resolve this problem. The drive of affectivity coming from the pre-individual means that the individual feels called back by something he perceives as external to himself, which no longer allows him to perceive himself as complete. Perception and affectivity, therefore, involve a *disparation*, that is, an incompatibility that must be resolved by the individual. Regarding *disparation*, Simondon writes in a footnote:

> this word is borrowed from the psycho-physiological theory of perception; there is disparation when two twin ensembles that are not completely superposable, such as the left retinal image and the right retinal image, are grasped together as a system, allowing for the formation of a single ensemble at a higher degree that integrates all their elements due to a new dimension (for example, in the case of vision, the layering of depths offield) (*ILFI*, p. 226).

Carrozzini in his commentary on *ILFI* adds that «Simondon notes that the system from which an individuation originates must be in a condition of non-unity and non-identity, and therefore in a state of disparation» [35, p. 25]. The concept of disparation was given particular prominence by Deleuze in his reinterpretation of Simondonian philosophy. In particular, in *Difference and Repetition* of 1968, Deleuze argues that «the act of individuation does not consist in suppressing the problem, but in integrating the elements of disparity in a state of coupling that ensures their internal resonance» [79, p. 395]. Resolution can only be achieved if the individual enters into a relationship with the collective by bringing about the emergence of emotion and action. The former incorporates the drive of affectivity and the latter that of perception/sensation. But the transductive movement that allows the problem of disparity to be resolved is the one that makes emotion and action complement each other, that is, the individualisation of the collective. Spirituality is represented by both emotion and action as two sides of a mountain that reach the same summit point. The spirituality expressed by emotion is that which resonates in the individual-subject and makes it capable of recognising itself in its being in continuous relation with the pre-individual and, therefore, a participant in eternity. That expressed by action, on the other hand, is the spirituality that emerges from the individual-subject in relation to the collective and is expressed in an eternity given to the world, for example to monuments or institutions or art in general.

In the absence of this three-way relationship, the individual-subject would be lost in the pre-individual to which it is drawn by affectivity and would fall into anxiety. The latter is felt when the individual-subject tries to solve the problem of the disparity apprehended through affectivity and perception without the intervention of the collective:

> the being is like its own object, but an object as important as itself; it could be said that the subject becomes object and witnesses its own expansion according to dimensions it cannot assume (*ILFI*, p. 283).

Anxiety thus consists, for Simondon, in the objectification of the individual-subject who loses all reference to his own interiority: he looks from the outside forgetting the inside. The energy charge of the pre-individual takes over from the individual, making the disparity ever deeper. The individual-subject perceives itself as increasingly detached from reality, but closer and closer to what it feels as wholeness, the pre-individual. This process cannot lead to the identification of the collective because it distances the individual-subject from both emotion and action. The only possible end point of anxiety is, according to Simondon, the annihilation of being as the abandonment of individuality in the direction of another potential individuation. It is as if the movement triggered by anxiety were an inverse path to ontogenesis, as when one watches a film being brought back and the relationships that had been woven together return to their previous state of isolation (*ILFI*, p. 284).

If, however, the three-way relationship between the pre-individual, the individuated and the collective is maintained, Simondon argues, a network of viewpoints and attitudes (given by the resonance between emotion and action) emerges in which the new individuation of the collective includes both the individuated's relationship to the world and that of the individuated to each other. This means that that of the collective or trans-individual is not merely an inter-individual communication, i.e. a relationship between individuals, but a true individualisation of a new level of the real.

In order for this new individuation to take place, what is already individuated must first engage in the process of individuating. For Simondon, *individualisation* is a psychic operation that extends itself by discovering meaning within a set of signals, thereby enabling access to a new level of individuation. Contrary to *Gestaltpsychology*—which regarded meaning as an original, pre-given form of reality—Simondon sees meaning as something that emerges from the encounter between an individualisation and an individuation (i.e., between two levels of reality). This encounter generates a transductive leap, which propels the process toward a new phase of individuation.

Unlike individuation, individualisation does not occur through transductive leaps but unfolds as a continuous flow, dividing the living being into distinct domains—such as thought and the body. What maintains coherence between individuation and individualisation is *personalisation*: the process by which personality traits are structured. These traits are not fixed, but function to hold together the singularity of individuation and the ongoing flux of individualisation, adapting or shifting as needed. Personality thus emerges as a stratification of successive structures, each replacing earlier configurations that can no longer reconcile the two dimensions—sometimes incorporating elements from outdated structures that still retain partial relevance.

In order to comprehend the distinction that Simondon makes between individuation, individualisation and personalisation, it is proposed that the following example is considered. Imagine a person who has worked for years as an engineer (a stable configuration of his personality), but who, over time, begins to feel a deep unease, an intolerance that he cannot clearly explain. This unease does not concern a technical problem, but an internal tension between what he is and what he is becoming.

The person continues to do their job, maintains their habits, relationships and social roles (*individualisation flow*). At a certain point, thanks to an event (*transductive leap*), they discover that they have a hidden talent or that they do not really want to do that job and change their way of being in the world (*new individuation*). In order to hold together what they were before with this new self, they must generate new habits and new personality traits that integrate with some old ones (*personalisation*).

In closing this third part devoted to psychic individuation, Simondon focuses on the need, which has emerged from the examination of the functioning of the psychic world, to overcome absolute substantivalism as well as absolute dynamism. In the former, an attempt had been made to identify the individual with the substance to the exclusion of the operation; in the latter, on the other hand, it was the becoming that overcame the structure/substance. In both conditions, it was the relationship that lost value (*ILFI*, pp. 324–325). If personality, as the transductive point of contact between individuation and individualisation, is a stratification, then the study of the individual in the field of psychology must be approached according to a logic for which substance and becoming are no longer opposing attributes. Thus, for Simondon, the psychological individual is made up of layers of the transductive order of succession (becoming) and others of the transductive order of simultaneity (substance).

Simondon then ventures another hypothesis, according to which the difference between biological reality and psychical reality lies in the fact that in the latter the regime of transductivity, which is its primary characteristic, constitutes the reciprocity between the order of simultaneous and successive. In biological reality, on the other hand, this reciprocity cannot occur, leaving the living individual with only the two separate orders: he can integrate past experience into the present but cannot do the reverse, i.e. he cannot reflect. From the outset of his written work, Simondon employs the term 'reflection' to denote the genuine philosophical attitude:

> A reflexive attitude must begin by avoiding the postulation of a determined belonging or end at the moment it begins to exist and seeks to define itself. A philosophy that accepted to be defined by a label such as 'christian', 'marxist', 'phenomenological' would find the negation of its philosophical nature in this initial determination [33, p. 17].

Again in the same text, *Note sur l'attitude réflexive*, dated 1955 that was to be part of *ILFI* Simondon specifies that:

> Reflection is a special case of the relationship between a problem and the various operations with which it can be solved thanks to the presence of a subject that is already constituted but still incompletely balanced. A perfectly complete subject, without any lack of unity, would not be able to think and reflect [33, p. 22].

For Simondon, then, reflection is the right philosophical attitude present in the human individual (a phase of being that is structured into subjects) as a particular case of the relationship between a problematic (the individual) and the different operations that can be activated to resolve it. There can only be reflection if the individual is incomplete, available to transformation. [33, pp. 17–25].

In *ILFI*, *reflection* is the ability to imagine the present experience before it occurs, as if it were already past; it is the possibility of bringing the order of the subsequent and the simultaneous on an equal footing:

> the past behavior conserves the characteristics of interiority that make it a behavior; it conserves a certain coefficient of presence; inversely, the present behavior, [...] is already an experience in advance (*ILFI*, p. 309).

The psychic individual would thus consist of this reflexive activity self-powered by the biological individuality from which the incompatibility to be resolved continues to come. The psychic individuality would thus stand at the threshold between physical and biological reality as a «transductive relation of the world and the self» (*ILFI*, p. 310). It would allow the (biological) self and the (physical) world to really communicate after their (pre-individual) bond has been broken and a different reality is needed that can establish compatibility. The reflexive capacity of the human being is what enables him to produce technical objects and see them as supports in his relationship with the world.

2.9 Forme, Information, Potentiels

An interesting summary of Simondoni's proposal for a new—necessary—development of psychology can be found in the 1960 text entitled *Forme, information et potentiels*. Included as a supplement to the main thesis in its latest edition, the complete edition of 2005 (*ILFI*, pp. 674–700), it is actually a text written on the occasion of a lecture given on 27 February 1960 at the *Société française de philosophie* in the presence of several illustrious names[18] from the French academy (Carrozzini 2020a, pp. 193–194). It was then published in the Bulletin of the society in the fourth issue of the same year (1960).

Gaston Berger, then president of the Société, whose task it was to present Simondon's paper, had declared that Simondon would present to the audience a synthesis of his research work, which would concern «neither a psychology of work, nor a sociology of man at work, nor even less a sociology of the use of manufactured objects» [1, p. 193], but the possibility of founding an axiomatics of the human sciences and psychology on the basis of an inventive operation, i.e. the questioning of the conceptual systems used up to that point in order to find new ones. This project would not have taken off without a new and proactive look at the technical-scientific world that would have provided the new concepts on which to work (*ILFI*, p. 674). In fact, it was precisely the study of the genesis and evolution of the technical object that allowed Simondon to trace a paradigm to interpret in a new way the relationship between the notions of form, information and energy potentials that are the title of this intervention (*ILFI*, p. 676).

Starting from the analysis carried out by Simondon during the 1950s, and specifically during the 1956 course on the stages of development of modern psychology, the aim of the 1960 conference is to trace a possible line of continuity between psychology and sociology that can be extended to all the human sciences. In particular, Simondon declares that he focuses on psychology and sociology because, as fields of research whose object of study is difficult to pin down, they are the disciplines that have undergone the most fragmentation over time.

In order to trace an axiomatic unity of these sciences, as occurred for those of nature through the electromagnetic theory of light elaborated by Maxwell in 1864 (Maxwell [96]), it is first of all necessary, according to Simondon, to renounce the idea that there exists in psychology a «psychological atom», i.e. an ultimate element to be studied in isolation, and in sociology a «human group», i.e. a group of groups (*ILFI*, p. 676). For both psychology and sociology «we are always at the level of correlations» (*ILFI*, p. 676).

To speak of the correlations common to the various human sciences, Simondon analyses three key terms that could be the foundation of the axiomatics of what he calls «human Science» (*ILFI*, p. 677). The analysis carried out in these first pages calls to mind the first chapter of the first part of *ILFI*, demonstrating the value of this essay as a possible synthesis of the work discussed in '58.

---

[18] Among those present were Jean Hyppoliye, Paul Ricouer, Gaston Berger and Jean Wahl.

The first concept analysed is that of form, a term that has been used since Plato to talk about the human question and that has undergone various interpretations up to the contemporary one elaborated by *Gestaltpsychologie*. In Plato, form is archetypal, in the sense that it exists before the reality we observe and is its foundation. The Platonic archetype is the first imprint given to reality by something that is eternal, immobile and unique. The Greek term used by Plato to denote form, in addition to ἀρχή, is τύπος meaning type, model, mould or matrix and whose origin belongs to the world of craftsmanship, particularly metalworking. It was originally used to denote the mould used to engrave a piece of metal that would become a coin. The τύπος was always the same and was stamped into every piece of metal. If it had not also been made of metal, it could not have struck the future coin to engrave a figure or inscription on it. What differentiates the τύπος from the different coins is the fact that the former, if it were lost, would have to be reproduced from a coin that could never be as perfect as the original mould because in the engraving operation there are several variables that could make the individual coin not exactly identical to the others. From this conviction Plato elaborates the theory of the archetypal form according to which random reality (individual coins) will never be an exact copy of the forms that originated it, but will tend to achieve that perfection.

Simondon states that in this theory, the process of interaction between the τύπος and the coins (the correlation) is irreversible since it is not possible to trace the model, the archetypal form, back to the coins; but it is also asymmetrical since the τύπος has an ontologically superior value to the coins and can subsist without the presence of the coins (*ILFI*, pp. 678–679). Archetypal forms do not change because they are perfect, and everything derived from them is instead subject to genesis and change. In this way, Simondon argues, Plato can find a universe in which perfection has always existed and man does nothing but mutate in order to achieve it, while the archetypal ideas or forms persist beyond the sky of fixed stars in perpetual preservation. What emerges from Plato's theory of form is a world of the mutable that must tend towards absolute immobility, like the ideal city described in the dialogue *Republic*, in which the philosopher-magistrate controls the citizens according to the pre-knowledge of each of them and the laws promulgated are designed to ensure that the organisation, the structure given to the city remains immutable and eternal.

Despite this, Simondon notes the fact that Plato, as reported in Books M and N of Aristotle's *Metaphysics*, does not seem to be completely satisfied with this theory since it is not really able to explain the sensible world. If ideas or archetypal forms are immobile, how can they generate something, since genesis is a process belonging to the world of becoming? It is through the theory of the indefinite Dyad[19] that Plato,

---

[19] Plato attempts to explain how ideas can generate the sensible world despite being immobile by explaining the principles that allow ideas to be multiple and at the same time remain unique. The first principle is that of the Dyad that acts on ideas by giving them the ability to imprint objects, reality, with the possibility of change. The second principle is that of the One that restrains the first principle from its multiplying capacity into structures that are able to maintain themselves without mutating.

in the *Timaeus*, attempts to answer this question, but the Platonic tradition that will reach the Middle Ages up to contemporary times will not particularly take this late revision into account (*ILFI*, p. 680).

In an attempt to resolve the question of the encounter between the immobility of unity and the becoming of the manifold, Aristotle proposes a theory of form that emphasises the genesis and transformation process of the individual. Simondon emphasises that this theory, unlike that of classical Platonism, defines a «horizontal relationship» between form and matter. Whereas in Plato form came before matter on both the ontological and gnoseological levels, in Aristotle the hylomorphic scheme envisages a meeting of the two within the συμβολον in which becoming is constitutive of individual being. Aristotle's challenge is to break out of the trap in which Plato had enclosed himself in which, in order to explain becoming, one had to refer to one or more principles that were divorced from becoming itself and that bestowed this capacity. The power of becoming, in Aristotle, is instead internal to the form-matter pair as the motor that drives the individual towards his *entelecheia*, his realisation through an evolutionary process. The hylomorphic scheme promoted by Aristotle, therefore, also opposes the Platonic thesis with respect to the gnoseological aspect; whereas for Plato the first per se was the archetypal form of being and—in the theory of *anamnesis*—also the first that originates and to which knowledge is maintained, in Aristotle it is the individual, the συμβολον, that becomes the primary source of being and knowledge.

Despite the attempt to find an explanation for becoming by stepping outside the need for an initial motor that does not move in turn, Aristotle seems to fall back into the Platonic error when he argues that in order for something to be known, it is necessary for at least one thing to be immovable. The latter is defined by Aristotle as ὑποκείμενον, i.e. *substratum*, that which lies beneath or performs a function. In fact, as recent critics have shown [80, 81], it is possible to mitigate what appears to be an internal contradiction in Aristotle's thesis by reading the *Physics* and the biological works that deal with movement in animals. In these works, Aristotle clarifies what the *substratum* is for him and that the necessity for something to be motionless in order for it to be perceived and known does not call for absolute immobility. In fact, the substratum is nothing more than something that performs the function of a pivot within a relation, and therefore its remaining still is primarily functional and therefore relative. Anything can be *substratum* and this can be understood from the fact that, for Aristotle, matter is the *substratum* of form, form is the *substratum* of accidents, and even the *synolus*, the individual, is the *substratum* of form and matter so that the latter stand together.

While not contemplating this reading of the Aristotelian sense of substratum as «articulation» or «pivot» (*arthron*) (Aristotle, *De Motu Animalium*) Simondon admits, somewhat mitigating the criticism contained in the chapter *Form and Matter* in the main thesis, that in Aristotelian theory form is an «eminently operative notion» (*ILFI*, p. 682) and, as such, could account for ontogenetic processes at least from the point of view of the individual. What it cannot explain, however, according to

Simondon, are the developmental processes of groups. Thus, Aristotle's notion of form does not turn out to be the one sought to serve as the foundation for an axiomatics of the human sciences.

Despite the continual search for a possible mediation between the Platonic and Aristotelian theses carried out by scholars in different eras, only the *Gestaltpsychologie* of the late nineteenth century seems to have achieved a first result in this direction. According to Simondon, through the notion of *field* that *Gestaltpsychologie* borrows from physics,[20] a new relationship of reciprocity can be established between unity and becoming as well as between the whole and the parts. In fact, what happens in any kind of field is that the parts/poles between which the field is generated can perform two different functions: that of receptors of the forces in the field and that of forces that create the field and can modify it. Through the concept of the field and the functions that take place within it, contemporary physics has reopened the internal discussion of ancient philosophy by bringing into play that resolution sought by Aristotle of an empty location (that of the pivot/article function), which can be occupied at any instant by different elements (*ILFI*, p. 683). What Aristotle seems to have lacked, despite the fact that he was the first to use the Greek term for it (ενέργεια), is the concept of energy as it is fundamental to field theory.

Gestaltpsychologie theory uses the concept of field to deal with perception (first with Brentano and then with Koehler, Koffka and Lewin), defining the perceptual field as a set of different elements that are within it and, at the same time, constitute it by being in continuous interaction. To perceive is to grasp a particular configuration of the entire perceptual field, hence, a way in which the different elements interact with each other. The notion of field is also extended by *Gestaltpsychologie* to the percipient subject who becomes a field-subject or internal field that stands in relation to the field-object or external field. Acting would therefore amount to finding a common configuration between these two fields as the discovery of a stable equilibrium point (the good form).

It is in this that lies, according to Simondon, the core of the problematic of *Gestaltpsychologie*: the fact of tracing a structure of structures that, insofar as it contains both the archetypal and the ilomorphic form, can only be perfect and stable (*ILFI*, p. 685). The good form or structure is that which drives the elements of the fields to the completion of the best configuration (Aristotelian *entelecheia*) and is also that which imposes its structure on all others as the most stable. However, this means, according to Simondon, that the good form is the most probable one since, according to the laws of thermodynamics, the natural and irreversible tendency of the universe moves in the direction of a stable equilibrium, i.e. towards a state in which no form of energy can be transformed any more (maximum entropy):

> *The most stable state is a state of death, i.e. a degraded state starting from which no transformation is possible without the intervention of an external energy to the degraded system* (*ILFI*, p. 685).

[20] Concept initially thought of by Faraday, but later confirmed and made explicit by Maxwell with his theory of electromagnetism.

As much as *Gestaltpsychologie* attempts to resolve the question of the encounter between the whole and the parts by pursuing the Greek ideal of stability, it seems to lose sight of one of the major discoveries of the nineteenth century, thus leaving the work of intersecting Platonic and Aristotelian theory incomplete. The good form of *Gestaltpsychologie*, in fact, cannot explain the evolutionary process, the becoming of reality because, being stable, the form pursued within the equilibrium of the systems it describes cannot change. Instead, in order to overcome the problem inherent in any theory of form that has unfolded over the ages, Simondon considers it necessary to have recourse to information theory (which develops precisely from the observations of thermodynamics). In particular, he recalls not only the classical information theory elaborated by Claude Shannon in 1948, but also the interpretations and revisions made by Norbert Wiener. What Shannon's and Wiener's views have in common is that information must be conceived as the path to a condition that is difficult to represent as stable equilibrium or perfect compatibility between two systems that enter into a communication relationship. Unlike the classical «form», in fact, information is a relationship between two systems that align themselves by generating communication based on an equilibrium that appears neither the «best» nor the most «probable» but as somehow the exact opposite of this. In the information relationship, in fact, if a receiver were only to align himself to a message he knows from the start, he would get no real information from the sender. In information exchange, on the other hand, the receiver is neither aware of nor able to predict everything that will come in the message. This is because, Simondon observes, no one would seek a message or send one if they were not attempting to know or realise something new, i.e. a new information condition as a novel state of equilibrium between two systems. Returning to the principles of thermodynamics, according to which the direction in which the entire universe is moving is towards maximum entropy (the state in which no transformation of energy will be possible except through an intervention outside the universe itself), information exchange seems to be a process that seeks to move in the opposite direction, towards what has been defined first by Erwin [82] and then by Léon [83, 84]) as *negative entropy* or *neghentropy*.

Simondon notes, however, the paradox that would arise if one were to use the information paradigm as elaborated by Shannon and Wiener within the field of psychology and sociology. The aim is to find a suitable paradigm to found a human science that can hold all disciplines together. In classical information theory, the engineers' goal is to trace and enable a correlation between a sender and a receiver; to do this, the amount of information required depends on how close or how weak the correlation between the two poles is:

> The operator merely requires a very small quantity of information from the emitter, i.e. from the object on which he works, from the machine he controls. The form would therefore be one that requires the least quantity of information (*ILFI*, p. 687).

In the psychosocial sphere, this thesis cannot be sufficient, since it is not enough to quantify information in order to obtain the best form, but the qualitative aspect of it must also be traced. Simondon's proposal is therefore to try to modify the purely quantitative classical information theory to bring in the qualitative aspect. What he seeks is in this sense an *informational tension (ILFI*, p. 687). This is the same *tension* that one imagines when one speaks of the difference in energy potential between the two poles of the reference source. An energy voltage that is too low or too high does not allow for the activation of an electrical system, and it is therefore necessary to regulate it so that there can be information exchange in the system, the passage of electrical energy from one pole to another. Simondon's idea is to think of good form as a network made up of correlation nodes, «something unique and multiple at the same time» (*ILFI*, p. 688) without falling into contradiction. A network, therefore, in which each correlation node would play the role of whole or element depending on the type of configuration the network is to take on. The good form would be a pool of potentials in which compatibility and reverberation coexist (*ILFI*, p. 688) and whose *prägnanz* does not lie in stability, but in the capacity to configure/structure by activating the different nodes of the network it traverses. This thesis, as opposed to the classical information thesis, does not define the poles of the relationship *ex ante* but makes them emerge from it. Consequently, Simondon states that the theory of information tension cannot be applied except from experience and, therefore, from ontogenetic processes, both individual and group, in which a pattern propagates itself in the space–time surrounding it, attempting to structure the energies present, that is, to configure them in such a way as to make them momentarily stable. This can occur if the environment in which the pattern or «crystalline germ» propagates is in a metastable state, i.e. in a condition that in chemistry is called *supersaturation*, in which there is an excess of energies that need to be released. In chemistry, the *supersaturation* is the state of a solution that contains more dissolved material than could be dissolved by the solvent under normal circumstances. Simondon takes this concept to describe the condition in which the pre-individual must be found in order to generate a process of individuation, but also the condition that allows any transductive transition from one order of magnitude to another. In note 14 of *ILFI* Simondon also states that «supersaturation actually constitutes a physico-chemical constraint, which, in turn, generates metastability» (p. 387). Thus supersaturation generates the condition of metastability of a system because the forces present are no longer in a state of compatibility. The crystalline germ intervenes in this supersaturated solution to bring it to a stable phase, releasing some of the energies present:

> This particular type of rapport that exists between the *structural germ's tension of information and the informable metastable domain*, which harbors a potential energy, turns the operation of form- taking into a *modulation* [...] (*ILFI*, pp. 689–690).

As is his custom, Simondon uses the technological analogy with the term *modulation*, indicating by it that same process that occurs through relays. This modulation process does not happen in an instant, but follows stages: the form moves towards regions that have not yet received configuration. This type of transformative movement is what Simondon has already called *transduction*, i.e. that transformation that proceeds by leaps and bounds, allowing the passage from one level of reality to another. However, the process of taking form cannot take place with any form in any field because there are a certain limited number of possible configurations for each field:

> In the transductive operation of modulation, which is veritably the hylomorphic operation, not just any form can unleash the actualization of potential energy of any metastable field whatsoever: the tension of a schema's form depends on the field to which it is applied (*ILFI*, p. 692).

Within a metastable field there is potential energy that can actualise itself, through the process of taking shape, only in some structures and not in others. The freedom to actualise energy potentials is therefore limited to the characteristics of the potentials themselves. Illuminating for understanding this thesis is the example described in the first chapter of the first part of *ILFI* when, in an attempt to demonstrate the unfoundedness of the Aristotelian hylomorphic theory, Simondon describes the process of taking shape of a brick. In this process, the clay of which the brick will be made is composed of molecules that virtually have the characteristic of being able to maintain themselves in a more liquid state (brute clay) and of being able to reject all the watery component to become more solid (*ILFI*, pp. 22–24). This so-called colloidal[21] characteristic of clay molecules allows it to be used to make bricks, but not to make other objects. In fact, the clay molecules have no informational tension in relation to any field other than that of the brick or any other field that does not require these specific characteristics of clay.

Returning to the '60 text, Simondon states that it must be possible to apply this kind of form-taking process to every level/stage of reality. In particular, the ontogenesis of a living organism occurs starting from the existence of a field of potential energies (body field, but also psychic and behavioural field) within which, starting from a metastable state in which the energies try to organise themselves, a *crystalline germ* or type of energy begins to extend itself within the field, organising it according to one of the possible patterns. Subsequently, the living organism continues its ontogenetic process through phases of adaptation and maladaptation in which specific characters of the field are either lost in order to acquire new ones or are preserved to become active again at a later stage in which they find a function within a new configuration (*ILFI*, p. 693).

Simondon extends this energetic theory of shape-taking to the field of perception and knowledge. In the case of visual perception, incompatibility or oversaturation of the system would occur if we had not developed the ability to perceive depth to compensate for the fact that the retinas of the two eyes do not receive exactly the same message. Through the acquisition of depth, we are instead able to integrate the two messages and avoid the so-called binocular *disparation*.

The Platonic and Aristotelian theses can therefore find their point of contact in the theory of energetic potentials since, on the one hand, the field would represent the Aristotelian matter that can receive a form, and, on the other hand, the crystalline germ would perform the function of the archetypal form, provided, however, that the field is in a supersaturated condition such as to trigger the process of taking form.

---

[21] A molecule can be called colloidal if it is in an intermediate state between a solution and a dispersion, i.e. if it contains both continuous dispersing substances and discontinuous microscopic substances. It is suitable for generating colloidal formulations, i.e. solutions in which the molecules are 88 nm to 1 μm in size such that they cannot be seen with the naked eye and can assume different conformations.

The germ would proceed by transforming the newly configured zone into a new modulator of potential energies for the part of the field yet to be configured. In this way, the germ would take on the same task as the Aristotelian ὑποκείμενον in the «articulatory» interpretation: it would move from time to time to make a different part of the field play the role of modulator/articulator.

Finally, Simondon wonders whether the same energetic theory can be applied to the social and psychosocial field. The supersaturated stage of the field would correspond to a pre-revolutionary social condition in which a more or less equally likely series of events could occur. The germ of form taking would be one of the events that takes hold of the other potentially verifiable events and begins to amplify by extending its configuration to the entire field (*ILFI*, p. 696).

Simondon's intervention concludes with two very important clarifications: the first concerns the distinction between field and domain; the second the opposition between the concept of maladaptation and that of degradation. As far as the first clarification is concerned, Simondon wants to distinguish the field as the place in which energy tensions are nested and from which the process of taking shape through the germ begins; while the domain is the portion of reality that can be configured through the transductive operation of taking shape.

The second clarification concerns, on the other hand, the need to connote the term maladaptation with a propositional aura: in this sense it is distinguished from degradation, which is nothing more than the process that leads to an increase in entropy (a state in which no other transformation is possible in the absence of external work) in a system, since on the contrary, maladaptation is the necessary condition for a transformation to occur (*ILFI*, p. 697).

Simondon's conclusion is that this energy theory could be the starting point for founding «an axiomatics of a human science» (*ILFI*, p. 697) that is not only based on a theory of form, but also on the notion of energy potentials.

References

1. Carrozzini G (2020) Variazioni su Simondon. Castelvecchi
2. Carrozzini G (2011) Gilbert Simondon filosofo della "mentalité technique". Mimesis/Centro Internazionale Insubrico
3. Barthélémy JH (2014) Simondon. Les Belles Lettres
4. Jalley E (2004) La crise de la psychologie à l'Université en France. L'Harmattan
5. Simondon G (2014) Sur la technique, 1953–1983. Puf
6. Carrozzini G (ed) (2008) Gilbert Simondon filosofo delle tecniche. Il Protagora, 36/12, Mimesis
7. Carrozzini G (2021) Gilbert Simondon. Del modo di esistenza di un individuo. Castelvecchi
8. Merleau-Ponty M (1968) L'union de l'âme et du corps chez Malebranche, Biran et Bergson. Notes prises au cours de M. Merleau- Ponty à l'École Normale Supérieure (1947–1948), recueillies et rédigées par Jean Deprun (Bibliothèque d'histoire de la philosophie). Vrin
9. Merleau-Ponty M (1959) Le philosophe et son ombre. In: Van Breda HL, Taminiaux J (eds) Edmund Husserl 1859–1959. Nijhoff
10. Merleau-Ponty M (1960) Èloge de la philosophie. Gallimard
11. Merleau-Ponty M (1964) Le visible et l'invisible. Gallimard
12. Moroncini B (2018) Corpo e Spazio: Husserl Merleau-Ponty, Lacan. Boll Filos 33:276–289

13. CCPLF (2018) Quel chemin la psychologie a-t-elle parcouru depuis 1920 avec Henri Piéron? Clin méditerranéennes 98(2):305–309
14. Parot F (2017) La psychologie française dans l'impasse. Du positivisme de Piéron au personnalisme de Fraisse. Éditions matériologiques
15. Piéron H (1908) L'évolution du psychisme. Revue du mois 291–310
16. Piéron H (1926) Intervention à la société de philosophie. Bulletin de la société française de philosophie 26(1):6–8
17. Hyppolite J (1964) En hommage à Gaston Bachelard: L'épistémologie de G. Bachelard. Rev D'Histoire Sci Leurs Appl 17(1):1–12
18. Bachelard G (1938) La psychanalyse du feu. Gallimard
19. Bachelard G (1942) L'Eau et les Rêves. Librairie José Corti
20. Bachelard G (1943) L'Air et les songes. Librairie José Corti
21. Bachelard G (1948) La Terre et les Rêveries de la volonté. Librairie José Corti
22. Simondon G (2017) On the mode of existence of technical objects. Univocal Publishing [MEOT]
23. Bachelard G (1940) La Philosophie du non. Puf
24. Le Ru V (2017) Bachelard, philosophe du jour, philosophe de la nuit. Diogène 257(1):41–52
25. Canguilhem G (1983) Dialectique et philosophie du non chez Gaston Bachelard, in Id., Études d'histoire et de philosophie des sciences. Vrin
26. Bontemps V (2019) Simondon et Bachelard. Implications philosophiques. https://www.implications-philosophiques.org/simondon-et-bachelard/
27. Fraisse P (1992) Autobiographie. In: Parot F, Richelle M (eds) Psychologues de langue française; autobiographies. Puf, pp 79–96
28. Fraisse P (1956) Les Structures rythmiques. Publications universitaires de Louvain
29. Fraisse P (1957) Psychologie du temps. Puf
30. Canguilhem G (2015) La connaissance de la vie. Vrin
31. Maggiore V (2018) Ernst Haeckel e il "rompicapo" dello sviluppo embrionale tra arte e scienza. S&F SCIENZAEFILOSOFIA.IT 20(18):258–276
32. Maggiore V (2020) Ernst Haeckel tra Estetica e Morfologia. Un pensiero che prende forma. QuiEdit
33. Simondon G (2016) Sur la philosophie 1950–1980. Puf
34. Piaget J (1924) Le jugement et le raisonnement chez l'enfant. Delachaux et Niestlé
35. Carrozzini G (2020b) Simondoniana. Mimesis
36. Simondon G (2015) Sur la psychologie, 1956–1967. Puf
37. Brentano F (2015) Psychology from an empirical standpoint. Routledge
38. Bergson H (1998) Creative Evolution. Dover Publications
39. Wiener N (1948) Cybernetics, or control and communication in the animal and the machine. MIT Press
40. Simondon G, Le Terrier F (1957) La Psychologie moderne. In: Daumas M Histoire de la science, Des origines au XX° siècle. Gallimard
41. Pugliese A (2009) Unicità e relazione. Intersoggettività, genesi e Io puro in Husserl. Mimesis
42. Stiegler B (1994) La maïeutique de l'objet comme organisation de l'inorganique. Gilbert Simondon: une pensée de l'individuation et de la technique. Albin Michel
43. Husserl E (1982) Ideas pertaining to a pure phenomenology and to a phenomenological philosophy. Springer
44. Rozzoni C, Bianchetti M (2004) Leib o Körper, chair e corpus: la filosofia e la nozione di corpo. Intervista a Mauro Carbone. Chora IV(9)
45. Merleau-Ponty M (2006) Vie et individuation. Chiasmi International 7 Vrin
46. Lisciani Pietrini E (2007) Fuori della persona. Sull'«impersonale»: Merleau-Ponty Bergson Deleuze. Filos Polit 3(21):393–409
47. Lisciani PE (2008) Oltre la persona. Merleau-Ponty e l'«impersonale», *Δα΄ιμων*. Rev de Filosofía 44:119–132
48. https://web.archive.org/web/20180313094130/http://childstudycenter.yale.edu/about/history.aspx

49. Le Moli A (2023) Technophysis. Le tecniche della natura, 2th edn. Palermo University Press
50. Le Roux R (2018) Une histoire de la cybernétique en France (1948–1975). Classiques Garnier
51. Couffignal L (1933) Les Machines à calculer, leurs principes, leur évolution. Gauthier-Villars
52. Couffignal L (1952) Les Machines à penser. Editions le Minuit
53. Canguilhem G (1937) Descartes et la technique. Travaux du IXe Congrès International de Philosophie 2:77–85
54. Lecourt D (2016) Georges Canguilhem. Puf
55. Canguilhem G (2005) Le normal et le pathologique. Puf
56. Canguilhem G (1955) La Formation du concept de réflexe aux XXVIIe et XXVIIIe siècles. Puf
57. Bachelard G (1934-1935) L'Idéalisme discursif. Recherches philosophiques 4:21–29
58. Poccia D (2011) "Altrove che nella verità?" Scienza, vita e filosofia in Georges Canguilhem. Nóema 2. https://doi.org/10.13130/2239-5474/1403
59. Nicolas S (2002) Histoire de la psychologie française. In Press, Naissance d'une nouvelle science
60. Laloy É (1971) Pour arraisonner le devenir: de nouveaux concepts forgés par Gilbert Simondon, *Travaux et Jours*, 7; (2002). L'enseignement philosophique 53(1):47–62
61. Watson JB (1919) Psychology from the standpoint of a behaviorist. J B Lippincott Company
62. AA VV (1994) *Gilbert Simondon. Une pensée de l'individuation et de la technique.* Albin Michel
63. Toccafondi F (2000) Il Tutto e le parti. La Gestaltpsychologie tra filosofia e ricerca sperimentale (1912–1922). Franco Angeli
64. Heisenberg W, Schrödinger E, Born M, Auger P (1961) On modern physics. Clarkson N. Potter
65. Guéroult M (ed) (1965) Le concept d'information dans la science contemporaine. Gauthier-Villars Les éditions de Minuit
66. Simondon G (1964–65) L'instinct. Bull Psychol 18(235):75–103
67. Simondon G (2010) Communication et information. Cours et conférences. La Transparence
68. Marx K (1867) Das Kapital. Verlag
69. Mumford L (1972) Art and technics. Columbia University Press
70. Adorno TW, Horkheimer M (1947) Dialektik der Aufklärung. Querido Verlag
71. Marcuse H (1964) One-dimensional man: studies in the ideology of advanced industrial society. Beacon Press
72. Anders G (1956) Die Antiquiertheit des Menschen. C.H. Beck
73. Simondon G (2006) Cours sur la perception (1964–1965). La Transparence
74. Calì C (2003) Percezione e qualità gestaltiche Saggio sulla scuola di Brentano. Rivista di estetica 22(1):183–241
75. Poggi S (2013) Herbart, Mach, Ehrenfels: alle origini della Gestaltpsychologie. In: Erle G (ed) (2013) Limite e l'infinito: studi in onore di Antonio Moretto. CLUEB, pp 235–248
76. Van Caneghem D (1989) Hommage à Gilbert Simondon. Bulletin de Psychologie 42(392):815–836
77. Lewin K (1936) Principles of topological psychology. McGraw-Hill Book Company
78. Balibar E (2002) Spinoza. Il transindividuale. Ghibli
79. Deleuze G (1968) Différence et répétition. Puf
80. Laspia P (1997) L'articolazione linguistica. Origini biologiche di una metafora. Carocci
81. Laspia P (2023) Logos e Physis. Palermo University Press, Il libro A della Physica di Aristotele
82. Schrödinger E (1944) What is life—the physical aspect of the living cell. Cambridge University Press
83. Brillouin L (1953) Negentropy principle of information. J Appl Phys 24(9):1152–1163
84. Brillouin L (1959) La science et la théorie de l'information. Masson
85. Deleuze G (1966) Gilbert Simondon, L'individu et sa genèse physico-biologique. Rev Philos de la Fr et de l'étranger 156(1–3)
86. Merleau-Ponty M (1942) La structure du comportement. Puf
87. Merleau-Ponty M (1945) Phénoménologie de la perception. Gallimard
88. Simondon G (1989) L'Individuation psychique et collective. Aubier [IPC]

89. Simondon G (2020) Individuation in light of notions of form and information. University of Minnesota Press [ILFI]
90. Smargiassi M (2009) Solipsismo e intersoggettività nella fenomenologia trascendentale di Edmund Husserl. Dialegesthai 11 https://purl.org/mdd/mario-smargiassi-04
91. Stiegler B (1996) La technique et le temps, II-La désorentation. Galilée
92. Stiegler B (1998) Temps et individuation technique, psychique et collective dans l'œuvre de Simondon. Intellectica 26–27:241–256
93. Bachelard G (1927), *Essai sur la connaissance approchée*, Vrin.
94. Friedmann G (1956), *Le travail en miettes: spécialisation et loisirs*, Gallimard.
95. VV AA (1990), *Cahiers philosophiques*, 43.
96. Maxwell (1865), *A Dynamical Theory of the Electromagnetic Field*, The Royal Society.

Chapter 3
The Courses of the 1960s–1970s

3.1 Introduction

Following the defense of his two doctoral theses and the lecture delivered at the Société française de philosophie in 1960, Gilbert Simondon dedicated the 1960s to an in-depth exploration of the themes outlined in the section of *L'Individuation à la lumière des notions de forme et d'information* (*ILFI* [14]) concerned with psychic individuation. During this period, he taught a series of courses—primarily in psychology but also in philosophy—at the universities of Poitiers, Lyon, and Paris, continuing at least until 1966. The first in-depth course, entitled *Psychosociologie de la technicité*, was taught at the University of Lyon between 1960 and 1961, while Simondon was still a professor at the Faculty of Arts in Poitiers. It was first published in the Bulletin de l'Ecole pratique de psychologie de Lyon at the end of 1960 and in 1961 [1, pp. 27–130]. Within these publications, the third part, entitled *Technicitè et sacralité*, is a lecture given by Simondon in 1961 in Bordeaux and later reported at the conclusion of the Lyon course.

In the first semester of the 1963–1964 academic year, Simondon delivered an untitled propaedeutic course at the University of Poitiers. The course transcripts, published in 2004 under the title *Deux leçons sur l'animal et l'homme* [2], provide key insights into the development of Simondon's theory of instinct, which would be more fully articulated the following year [10].

Both the course on instinct, of which only the transcription of the first part can be read in the Bulletin de psychologie, later published in volume 2015 [3, pp. 247–374], and the course on perception held at the Sorbonne, first published in the same journal and later published by La Transparence in 2006, date from 1964. In particular, the course on perception, which also occupied the semester of 1965 [11], is presented as a course halfway between the history of philosophy, to which the first part of *La perception dans la pensée occidentale* is devoted, and psychosociology, to which the

F. Sunseri, *Psychology and Technology in Gilbert Simondon*, UNIPA Springer Series,
https://doi.org/10.1007/978-3-032-25740-6_3

other four parts are devoted: *Rôle et sens biologique de la fonction perceptive, Perception et information: les effets psychologiques, Perception et affectivité, Perception et activité*.

At the turn of the year 1965–1966, Simondon gave a course entitled *Imagination and Invention* (*IEI* [4]) at the Institute of Psychology at the Sorbonne, parts of which were also published in six different issues of the Bulletin de psychologie and then republished in the 2008 edition of La Transparence [4]. Because of its particular role in the assonances found between the cycle of images described, *ILFI*'s process of individuation and *MEOT's* materialization, the next section will be devoted to the analysis of this course.

Between 1966 and 1967, also at the Sorbonne, Simondon taught two courses: *La sensibilité and Initiation à la psychologie moderne.* Both were published in several issues of the Bulletin de psychologie between 1966 and 1967.

Between 1968 and 1970 he explored perception in depth in *Perception et modulation* and *La perception de longue durée*, and communication in *Cours sur la communication.* All three were published by Puf: the first and third in the 2010 anthology *Communication et information* ([12] pp. 187–246; pp. 53–156) and the second in *La résolution des problèmes* in 2018 [5].

In 1974–75-76, Simondon returned to the theme of the complex relationship between man and technical objects with three courses: *L'homme et l'object, La résolution des problèmes* and *Invention et créativité*. These courses were also taught for the general psychology department at the Sorbonne, and the typewritten notes were published for the first time in 2018 in *La résolution des problems* [5].

To illustrate how Simondon's psychological research fits into the overall project of his allagmatics and his philosophy of individuation in general, the second chapter of our research will therefore analyse the course on imagination and invention as a key point of his psycho-sociological perspective. This will be done to demonstrate, through the lens of a particular didactic experience, the profoundly interdisciplinary direction that Simondonian research insists on following the defence of the theses, which are thus the germ of a much broader organization. A later stage of our work will be devoted to the analysis of the entire body of Simondon's psychological works, to define their specific place within the *corpus.*

3.2 Imagination and Invention 1965–1966

Between 1965 and 1966 Simondon taught the general psychology course Imagination et Invention at the Institute of Psychology at the Sorbonne. As Jean-Yves Chateau reported in the *Introduction* to the French edition: «This course, then, is a final testament to what could be done in the General Psychology classroom toward the end of this period, when general psychology constituted an important element in the formation of philosophers» (*IEI* [4], p. IX). This is therefore an archaeological find of what was, until 1966, the teaching of general psychology in French universities: a compulsory course for both psychology and philosophy students. In fact, from the following academic year, philosophy students would no longer have any compulsory psychology course. The aim of this course (as stated by Simondon in the introduction) is to present a theory of the image which shows that aspects of it are «stages of a single activity undergoing a developmental process» (*IEI* [4], p. 3). As can be seen from the table of contents, Simondon believes that the image follows a process that, like individuation in its various stages, is never final. In fact, like the individual in *ILFI*, the *IEI* image proceeds to find a possible equilibrium with its associated environment until it arrives at an adequate organization, at least until a new germ (a new image) arrives to reactivate the process of imagination. The latter is not only the activity of producing and evoking images, but also of receiving them and giving them meaning. It is therefore a philosophical as well as a psycho-sociological task to study imagination.

In the introduction, Simondon states that the image is neither something wholly produced by the subject nor something wholly external as an object. Images contain a certain *opacity* that makes them *quasi-organisms* or *parasites* that inhabit the organism-subject while remaining halfway between subject and object (*IEI* [4], p. 9). On the one hand, images can activate a split in the subject that had found its equilibrium; on the other hand, they have the possibility of introducing new potential energy to solve problems.

It is precisely because of this *dual nature* (*dédoublement*) of the image that it is conceived by Simondon as an affective-emotional anticipation that prompts action (*semi-concrete*), but at the same time as not *prägnanz* enough to allow the subject to choose freely (*semi-abstract*). In this sense, the image acts as a regulator between the impulse to act provided by perception and the refusal to act provided by mere abstract thought:

> Any image is susceptible to incorporation into a process of materializing or idealizing recurrence; deposited in fashion, art, monuments, technical objects, the image becomes a source of complex perceptions triggering movements, cognitive representations, affections, and emotions. Almost all man-made objects are in some way image-objects; they bear latent meanings, not merely cognitive but conative and affective-emotional; image-objects are quasi-organisms or at least seeds capable of being reborn and developing in the subject. Even outside of the subject, they multiply through group exchanges and activities, propagating and reproducing themselves in a neotenic state until they find the opportunity to become reabsorbed and redeployed until they reach an imaginal phase in which they are reincorporated into a new invention (*IEI* [4], p. 13).

In these lines of *IEI*, one cannot fail to detect an echo of *ILFI* and not trace in the process of the image that circular causality proper to cybernetics. From the abstract or semi-abstract stage, which Simondon calls *neotenic*, the image begins to structure itself, becoming more and more semi-concrete, until it begins to spread outside the subject (*symbolisation*).

For Simondon, the image is not only a middle ground between subject and object, between abstract and concrete, but also something in which past and future coexist. In fact, the image can tend to recall the past by becoming a substitute for sensation, but it can also be the basis for prediction or anticipation by linking itself to the future. This dual nature of the image is always present: «the memory-image seeks to reincarnate and perpetuate itself, it carries a subjacent anticipation [...]. Anticipation, in turn, reprises old dreams and contains the echo of old aspirations» (*IEI* [4], p. 16). Simondon also finds the same dual nature in the images structured in collective life: they are on the one hand perspective extrapolations and on the other inventions.

Now, Simondon argues that ontogenetic studies (reported in *ILFI*) demonstrate the displacement that exists between the development of the different parts of a living being. In fact, not all organs develop at the same time and each phase of development is followed by a phase shift in the sense of a loss of the balance found and a reorganization. If this is the case for the ontogeny of living beings and if imagination is an activity of living beings, Simondon wonders whether a functional analogy cannot be established between ontogeny and imagination by thinking that there is a genetic dynamism in the latter as well, distinguished in stages.

Simondon's attempt is thus to try to distinguish the three stages of this genetic dynamism. It does not therefore seem incorrect to state that Simondon theorises imagination, like the realization in *MEOT*, as another mode of individuation. This attempt is also supported by an observation Simondon makes in the last paragraph of the *Introduction*: the genetic cycle of images can also be found in non-mental images, i.e. in culture, in the rhythms of life (day/night, the seasons) and in the becoming of civilizations (*IEI* [4], pp. 24–28).

The stages of imagination, for Simondon, are three: that of «pure and spontaneous growth» or anticipation; that of functional «apprenticeship» or experience; and that of «affective-emotional resonance» or «systematization» (*IEI* [4], pp. 18–19). In this process in stages, the image goes from being an anticipation not controlled by experience to becoming a mode of receiving information from the external environment and responding to this *input*, until it is transformed into a system, called the «mental world» (*IEI* [4], pp. 18–19) by Simondon. which reproduces the external environment by analogy. At the end of the process of imagination, therefore, there is a system of images that allow the subject to interact in a stable way with the external world. This stability is interrupted at the moment when a «*change of organization*» occurs to the images that had reached the third stage, and it is necessary to start the genesis process again. This change is called *invention* by Simondon: «a rebirth of the cycle of images, which makes it possible to face the environment with new anticipations from which adaptations emerge that had not been possible with the primitive anticipations» (*IEI* [4], pp. 18–19).

It is therefore necessary to retrace this process of the genesis of the image or imagination to understand what analogies are present between it and individuation. Moreover, this retracing will say much about the importance Simondon gives to the mental world as a regulator between the individual and the environment.

3.2.1 Motor Anticipation

The first stage of the imagination is the one dominated by the a priori image, i.e. by motor anticipation, which is linked to the fact that motor activity develops before perception, as was pointed out in the third part of *ILFI* (pp. 257–258). To support this claim, Simondon cites several biological studies, including those by the zoologist and geneticist Herbert Spencer Jennings. According to these studies, carried out on simple organisms (particularly protozoa), responses (the stimulus–response pattern) are preceded by spontaneous motor activity that is not linked to a specific perception. This shows that even in the simplest organisms, motor activity is activated before perception. Therefore, Simondon considers it possible that the image cycle in the most complex organisms (those that develop a nervous system) also follows the same process, starting with the motor aspect (*IEI* [4], pp. 30–31). This is an evolutionary inheritance that is preserved even as the nervous system develops and implies that every perception is preceded by «non-purposeful movement sketches» that are contained in the nervous system as possible adoptable behavioural patterns.

Simondon calls these non-purposeful motor activities *anticipations* and describes them as «organized initiative», i.e. having a form, or as patterns of behaviour (*IEI* [4], p. 32). In every organism there is a set of behavioral patterns that have the function of anticipating possible actions to be performed by the organism. The organism can also perform these actions in a vacuum, i.e. without being directed at an actual object.

In any case, the origin of these actions are the anticipations that the being possesses as a «program to unfold» (*IEI* [4], p. 33). Lorenz and Tinbergen's studies on hereditary coordination and Coghill and Carmichael's studies on embryology also show that «motor anticipation of behaviour occurs by way of endogenous development» and that the set of anticipations does not follow the stimulus–response law arising from perception and encounter with the environment (*IEI* [4], pp. 33–35). Children's play can also be traced back to this type of anticipatory image: the doll, like the toy train, represents another image that the child anticipates through motor activity, without being aware of the reference:

> All that is required is a basic and almost random object: a teddy bear, a rag ball, a pet; what matters is that the object can be cradled, rocked, and carried along; it is the support and aim of preformed movements constituting a "pattern" (*IEI* [4], p. 36).

Drawing on Merleau-Ponty's concept of the «bodily schema», Simondon proposes to think of these anticipations as what structures the bodily schema, i.e. the representation that each person makes of his or her own body. This schema includes the system of motor anticipations and is evidenced by the fact that from birth the organism can use its own organs. Simondon maintains that this capacity for motor anticipation can only be primary, since to perceive the movement of an object concretely, the individual must put himself in the position of the object. What develops is an *implicit animism* that projects the movement of the organism onto the environment (*IEI* [4], pp. 40–41).

The same kind of anticipatory image occurs, according to Simondon, in states of expectation at the *psychic* level. Simondon points out that the term "psychic" could be misunderstood, so he prefers to clarify its meaning: «a functioning of the organism that does not involve the organism as a whole in a situation» (*IEI* [4], pp. 40–41). From this definition we can argue that the psychic level of anticipation always comes after the biological level. For psychic anticipation to be structured, the subject must already have explored the environment (through the primary action schemata), which becomes the *territory* of action.

Simondon gives several examples to explain what a psychic level anticipation is. The first is that of phobias, which are not a consequence of a perceptual experience but «a mental activity of pre-perceptual anticipation» that amplifies the motor pattern of the primary or biological level. In fact, the motor anticipations of the biological level are magnified by the psyche and projected onto objects that do not really have those characteristics. Simondon gives the example of the phobia of poison that the individual acquires as an action pattern and subsequently projects onto other living beings (e.g. the toad) that do not actually have this power (*IEI* [4], p. 43).

In some cases, these projected anticipations are not completely realized as the motor anticipation is amplified even if no projection appears. This condition is, according to Simondon, that of compulsive behavior applied to repel motor anticipation that is amplified but fails to attach to a specific object.

Another case of projected anticipation is duplication, which occurs especially when biological anticipation concerns death. This fear leads the individual to duplicate his own image: he sees himself both as a corpse and as a subject mourning his own death. This duplication entails an anticipation of pain and man, unlike other animals, experiences it as something from which no escape action can save him: the duplication of self is the projection of one's own figure and actions onto a more powerful god (*IEI* [4], p. 46). In this process of duplication, man distinguishes between an inner world in which he feels protected and an outer world of which he is afraid: «the point of departure is the defensive gesture separating the near from the far, installing defenses to preserve the near, doubling in some fashion the subject [...] to fight the threatening adversity in the outside camp» (*IEI* [4], p. 47). Simondon traces the beginning of alienation and of all dualisms in this passage, which seems to call to mind the third part of *MEOT* and the mismatch between the magical and the techno-religious worlds. On the one hand the individual amplifies motor images onto objects and on the other he projects them onto more powerful entities that can save him from the adversities he encounters.

The last case of anticipatory images is that composed of those arising from mixed states of fear and hope. In these the amplification is neither totally directed outwards (the objects) nor inwards (the deities). There is a movement towards the center where the image remains immobilized at an intermediate distance between pure transcendence and pure immanence. Simondon believes that to speak of this middle place one could use Edgar Morin's expression: «third reality» (*IEI* [4], p. 49). In this third place are located, according to Simondon, the famous characters who are narrated at the same time as extremely human and like ordinary people and as belonging to the glossy world impossible to reach. This is the narrative that radio and television make of them, themselves intermediate spaces between reality and subject. The intermediate feeling between fear and hope is called 'wonder' by Simondon and can be linked to both subjects and objects. The latter become symbol-objects or *analogon* and, therefore, do not undergo any metamorphosis but *stand for* something else, for another reality (*IEI* [4], p. 51). For Simondon, wonder is not the only means by which the human individual can access this type of anticipation: «there are also productive forces which, more modestly, may construct an exclusive world in which the motor schema exerts itself and becomes concretized: it is the work of the amateur, that is, those who act out of love for what they do» (*IEI* [4], p. 54). Once again, in these pages, Simondon is very attentive to what is happening in his society and what he observed during his trip to the United States in the 1950s. Specifically, he uses "bricolage" as an example of this amateur work: a reproductive imagination that organises a certain permanent operational availability of materials and tools. The contemporary man who practices bricolage in his spare time is a disinterested craftsman who does not use anticipatory imagery to enclose himself in a fictitious world, but to access the realization of potentials (*IEI* [4], p. 55). What does not escape Simondon's gaze is the type of machines that are manufactured and sold to *bricoleurs*: all-rounder machines that do not, however, do everything simultaneously, i.e. machines that push heavily on virtuality. In this, according to Simondon, lies the clue to the imaginary anticipatory nature of bricolage (*IEI* [4], p. 55).

Finally, Simondon devotes a few pages to the possibility that the anticipatory image can act as a *principle of reflective knowledge*. Imaginary anticipation, in fact, can open reflection that is based on ideas, that is, on a priori images, and not on concepts. As in Plato's myth of the cave, ideas do not project themselves, but need a source of illumination which, in Plato's doctrine, is identified with the Good. Simondon points out that his reading of the Platonic theory of knowledge might seem paradoxical since the latter has always been considered a contemplative thesis. Instead, Simondon traces in Plato's theory of knowledge the «expression of a primordial image of pure movement» (*IEI* [4], p. 57). It is in the process of degradation from the archetype to the different images that the meaning of this expression is grasped as that which allows the projection to be grasped as it is unfolding. In fact, the contemplation of the ideas of Platonic theory would be an insight into the fact that movement is the origin of everything, that is, the transiting of all projection into existence (*IEI* [4], p. 58).

In Plotinus' contemplation and Bergson's *élan vital* he also finds the same grasp of original movement. In Bergson, however:

> The unity of the primordial upswelling is preserved in the continuity of life's movement that becomes diversified through matter; matter itself is like a movement that comes undone; the intuition of the moving is no longer constrained by being absorbed in the pure anticipation of the a priori creative gesture (*IEI* [4], p. 59).

Bergson's *vital impulse* would never be extinguished, continuing to create through movement. The degradation present in Plato and Plotinus would no longer exist in Bergson because the subject who knows participates in creation itself, has within him the same vital impulse that moves what is outside him. The subject would thus be able to carry further and anticipate at the same time the evolution of which he is also the object.

3.2.2 *The Infra-Perceptual Image*

The second part of the course is devoted to the second stage of the imaginative process: that linked to perception and experience. If in the anticipatory process the image is generated from a relationship between the subject and a yet unorganized environment, in this second stage the *infra-perceptual* image is generated from the relationship between the subject and the *territory*. The latter is «a world where there is no longer any novelty» (*IEI* [4], p. 63) and in which the living subject can carry out psychic activity, defined by Simondon as *secondary* in opposition to the primary or primitive activity of the first stage. In this passage, the distinction of degree (and not of essence) between other animals and the human animal is taken up. The second stage of imagination is linked to the level of development of the nervous system: the more developed it is, the greater the speed and capacity to organize its surroundings. This means, for Simondon, that even non-human animals can have psychic capacities if they are within their territory and if the latter is not too vast (*IEI* [4], p. 64). In this sense, humans, who have a more advanced nervous system, can organize an even larger environment faster, making their psychic abilities apparently more developed.

Now Simondon notes that experiments in ethology have shown that not all perceptual behaviour can wait for all information to be picked up by the subject. In some situations, it is necessary to act by conjecturing about the object in front of you. This is the first stage of transition from motor anticipation to perception and Simondon calls it *perceptual-motor behaviour* or as in 1964–65 set of *instinctual behaviours*. The subject receives a series of information from the object that triggers a chain reaction: each information flow corresponds to a response that opens the way for a new flow (*IEI* [4], p. 65).[1] At this stage, it is not possible to avoid the errors that are needed to make the image that is being generated sharper and sharper. These infra-perceptual images are still a priori because they do not come after the experience and do not provide a fully identifiable object. Simondon argues that they «prolong long-term anticipatory images» and can be defined as *gestalt-like* (*IEI* [4], p. 66) because they only capture certain aspects of the object being perceived such as posture, colour or shape without being able to obtain a complete image of it. These primary infra-perceptual images can, therefore, activate instinctual behaviour and select information that continues to reach the subject (*IEI* [4], pp. 72–73).

Infra-perceptual images therefore play a role in the acquisition of information through selection or differentiation. When a subject meets an object whose properties he already knows, the infra-perceptual image helps him to distinguish the new from the familiar in the stream of information that reaches him. In this sense, according to Simondon, the image is a *pattern*, a model, from which the subject proceeds in the identification, differentiation and, finally, derivation of the perceived object. The first step, that of identification, is based on the principle of constancy, which allows the subject to hold on to the object even though it is constantly changing its relationship both to the subject and to the environment/territory:

[1] Simondon calls him a *releaser*, a technical term for releasing, letting go.

> The perceptual function of constancy involves the deployment of an activity of short-term anticipation akin to that of servomechanisms and [motion] predictors ensuring the automatic tracking of objects [...] Constancy in the usual sense of the term is a particular case of short-term anticipation activity that allows for the identification of the object and the permanent reception of signals which allow for its tracking (*IEI* [4], p. 76).

In this passage, Simondon uses a functional analogy between the «behaviour» of a technical object (the servomechanisms or the predictor) and the perceptual image of constancy in order to make it clear what the function of the latter is. This type of image therefore includes the possibility of the object of perception transforming or changing position depending on the context.

Once the information flow generated by the object stabilizes, the subject perceives a figure against a background. This is the step of differentiation in which the subject abstracts the figure of the object from the different states of *compossibility* in which it could have been:

> It also appears in technics and more generally in the intuition of a knower sufficiently familiar with an organized reality to know concretely the compossibility of all its states [...] like the mother's intuition that her child is becoming sick; a mountaineer can better perceive the state that precedes a storm or a snowfall [...] This kind of perception of a state requires a special knowledge of all the details of a place (*IEI* [4], pp. 77-78).

The subject, therefore, must know the territory he is in to perceive through intuition. According to Simondon, Pascal had already given a name to this step of perceptive differentiation: *l'esprit de finesse* or *du cœur* (*IEI* [4], p. 77). The latter would be activated, according to Pascal, whenever a problem cannot be solved by the *esprit de géometrie*, i.e. by calculation. Then, Simondon argues, the differential infra-perceptual image intervenes, allowing the subject to have an idea of the object he is dealing with without having scientific/mathematical knowledge of it. This would explain, according to Simondon, the enormous amount of information flows a subject receives against its learning capacity limited to one binary unit per second (*IEI* [4], p. 79).[2]

The last step, that of derivation, occurs when the subject becomes accustomed to the perception, he or she is having because of the regularity of the information data. Only a strong detachment from the model built in the previous steps can trigger a new process of object organization. As long as the information data received by the subject remains regular enough, «the internal image of rhythms and regularities neutralizes the signals» and *habit* is generated (*IEI* [4], p. 80).

Once the object has become a perceptual habit, the subject's tendencies can alter its perception. Simondon reports Lipps' theory of empathy (*Einfühlung*) according to which a square placed in a stack of overlapping triangles and squares seems larger than its actual size because the subject identifies it as the object that holds the weight of all the others. According to Simondon, one should speak in this case of *implicit animism* or *latent organicism* of perception, since if the *Einfühlung* effect often occurs, it is because internal tendencies of objects are presupposed (*IEI* [4], p. 82).

---

[2] In this passage, Simondon's knowledge of Claude Shannon's classical information theory is once again evident.

The subject of perception, therefore, grasps the same properties that it possesses in the object it is perceiving: «to perceive a rope as straight because it is pulled by a force means to place oneself as an organism in the place of this rope, to make the rope an extension of the arm» (*IEI* [4], p. 83).

Now Simondon does not want to argue that perception is something stable, as *Gestaltpsychologie* had done before him, but to demonstrate that the process of imagination (like that of individuation and concretization) proceeds by metastable balances (*IEI* [4], p. 84). One of the proofs demonstrating the metastable character of perceptual equilibria is the reversibility of any previously established infra-perceptual images. If we take as an example those objects in which the gaze can focus on one figure or another, the state of pure stability would be, for Simondon, that in which both configurations overlap one another. In fact, Simondon argues in very few lines that the principle of stability of perception thought up by *Gestaltpsychologie* cannot be correct, since if infra-perceptual images remained irreversible, what would be obtained would be a condition of maximum degradation, i.e. a state in which no transformation is possible any longer (*IEI* [4], p. 84).[3]

Instead, Simondon offers the possibility of the isomorphism principle of *Gestaltpsychology*. The infra-perceptual image arises from the (always metastable) correspondence or compatibility between the molecular and molar levels of reality (*IEI* [4], p. 85). It is a balance that the infra-perceptual image finds to solve a problem, to mediate between a very large configuration and an elementary matter, between different orders of magnitude.[4]

Another concept from *Gestaltpsychology*, *prägnanz*, is overturned by Simondon at the end of Part 2. He argues that the poignancy of forms expresses the inherence of infra-perceptual images, which, as mentioned above, is not the result of a stable equilibrium. In perception, singularity is more *prägnanz* than regularity, even when it is a regularity that is thought to attract attention. An example given by Simondon is that of the «fairy chimneys», regular geometric shapes created by erosion. These chimneys are noticed because it is surprising to find such exact geometric shapes in nature where they are not normally found (*IEI* [4], p. 86). Simondon also argues that the poignancy of geometric forms can also be found in the case where they turn out to support other, more extreme forms (*IEI* [4], p. 87).[5]

Finally, the infra-perceptual image expresses the balance between the living and the environment that has now become territory, their coupling at a *key point* (*MEOT*, third part) that remains equidistant from the two systems at play. It is, therefore, a key with which the creature gains access to the world (*IEI* [4], p. 92).

---

[3] In this passage, Simondon does not refer directly to the theory of thermodynamics, but in several passages, he refers to «degradation», «disordered configuration» and «minimum potential energy of the system», all concepts that seem to recall entropy.

[4] The infra-perceptual image, therefore, appears to have transductive power. Simondon does not use this term in these pages, but he had already defined transduction as the ability to mediate between different orders of magnitude.

[5] Simondon gives the example of the round arch of Orléans Cathedral which is «surmounted by a low arch and framed by two ogive windows». The round arch supports the other structures and therefore its presence is significant.

3.2.3 The Image-Symbol

The third part of the course deals with the affective-emotional content of the image, i.e. a posteriori image or symbol. The first stage of this type of image is profoundly related to the balance found through the infra-perceptual image. One of the best-known examples is *imprinting* or *Prägung*. Studied and discovered by Lorenz in 1935, imprinting is «a mode of acquisition consisting of a kind of very brief and early impregnation» (*IEI* [4], p. 93). These behaviours are often mistaken for instincts but as mentioned earlier, instincts come into play in the image cycle in the motor phase. *Imprinting*, on the other hand, is something that is instilled in the individual from an exchange with the environment, which has now become territory, in the very first moments after birth. In fact, precisely because of its early character, Simondon deduces, imprinting is mistaken for an innate character of the personality of the individual or species (*IEI* [4], p. 94). This type of early reaction has an affective-emotional character, as has been emphasized by studies in psychoanalysis on the relationship between the child, its mother and its environment. What happens with the initial imprinting between child and mother is the incorporation of the parental figure into the ego, forming the basis for the various subsequent emotional-affective reactions. An image of the mother is constituted that will serve as a model for every type of affective-emotional reaction and if this image is not structured at a certain moment in the child's growth, the latter will not be able to build it later.

Imprinting, says Simondon, is not only carried out with living individuals, but also with objects. And at an early stage of development, it is not necessarily the case that the individual experiencing the imprinting is able to distinguish a living individual (the mother for example) from the object (the toy) (*IEI* [4], p. 96).

As with the previous stages, Simondon premises the discussion of the presence of certain images in humans with evidence of the same presence in other animal species. Regarding imprinting with objects, glaring examples are already present in birds and their relationship with the nest (*IEI* [4], p. 96).

In humans, as noted by Maria Montessori, the relationship with the object in the first years of life is just as important as the relationship with adult figures. In fact, Montessori, in her *Casa dei bambini*, places a whole series of objects within the child's reach, allowing them to manipulate them, to grasp them, thus generating the structuring of their key points (*IEI* [4], p. 97). But Montessori theory and practice derive, according to Simondon, from imprinting with objects that the child uses every day: toys. In them the child sees the «I» in the environment: «it forms a couple with the child, like the *symbolon* with the other *symbolon*, half of the primal whole from whence come *symbola* in a random division» (*IEI* [4], p. 98).[6] This does not mean that the toy becomes the duplicate of the child in the environment, but represents the other, «his best friend» (*IEI* [4], p. 99). In this way, the child can realize that in the environment there is not only the parental figure, but also the other with whom he has to deal, but in whom he also recognizes a certain resonance of himself. This image of

---

[6] One could read into this the magical phase present in *MEOT* or the pre-individual in *ILFI* from which the two *symbolae* are randomly divided.

impregnation, according to Simondon, cannot develop in adulthood because at that point the toy can no longer play the role of a key point as it is already socialized, and part of a cultural system made up of economic relations and rituals. What may happen in adulthood is that, if during childhood the child has developed an intense impregnation with the toy of choice, the affective-emotional effects may persist even when he becomes an adult (*IEI* [4], p. 100).

If in the case of imprinting the a posteriori generated image still has a strong link with the sensory stimulation, it is in the consecutive and then in the immediate image that this link becomes less and less tight. In the latter two cases, in fact, sensory stimulation plays a role even if the object that produced it is no longer present (*IEI* [4], p. 100). A case of consecutive imagery is Hering's: after the eye is stimulated by a very intense light, the bright spot appears to reappear shortly after the light has disappeared. According to Simondon, these images are «*persistence* phenomena», which, however, can immediately lose their persistence if another event occurs (*IEI* [4], p. 103).

A greater persistence occurs, on the other hand, with immediate images, which rather than being linked to a single sensory data are associated with perceptions as the result of a repetition and structuring of several sensory data. In the case where the sense involved is sight, one can speak more precisely, according to Simondon, of eidetic images: «a word that was seen in writing may be spelled backwards; numbers may be subjected to operations as if they were written on a blackboard» (*IEI* [4], p. 105). Simondon presents as an example of an eidetic image the one used by prodigy calculators or chess players with their eyes closed who manage to win a game by (mentally) maintaining the image of the chessboard. Simondon argues that eidetic images have a primitive or *wild* nature, but this does not make them a sign of regression: «eidetism may well appear to be one of the roots of artistic imagination and perhaps inventive imagination in general» (*IEI* [4], p. 107). In this passage, he opposes the psychologist and sociologist Daniel Essertier, who in *Les formes inférieures de l'explication* (1927) affermines that eidetic images have a primitive nature as they appear to be the residue of a capacity of primitive humans to use memory to survive. According to Simondon, this hypothesis starts from a misconception of evolution as a single progressive line.

Unlike the specifically visual eidetic image, the general immediate image is much more like a perception because concrete characters remain in it even if they are devoid of meaning. The eidetic image, on the other hand, preserves the concreteness of characters and gives them meaning by making them key points between the subject and the environment (*IEI* [4], p. 108). Furthermore, eidetic images are plastic because the subject can mentally modify them. To avoid any misunderstanding, Simondon specifies that the eidetic image is not, however, comparable to a photograph because what it retains of the characters of perception is only the essentials, while the details are lost. In addition, Simondon finds a continuity between imprinting and eidetic images in that starting from a subject/object that generates a certain attention, it is possible to foster both imprinting and the acquisition of these images (*IEI* [4], p. 109).

The link with sensory stimulation is absent in the case of memory-images, that allow the perceptual situation to which they refer to reappear as a representation even after it has been experienced for a long time. Although the eidetic image can act as an «active centre» for the memory-image, Simondon argues that there is a difference between them. There can be different types of memory-images,[7] depending on which sense is most involved (*IEI* [4], pp. 111–117). These images turn out to be the generalization of a particular perception. Simondon seems to be defending, if not endorsing, the empiricist thesis, found in Aristotle and opposed to idealism, that knowledge can only progress from experience through generalization. Simondon quotes Aristotle's phrase «the soul never thinks without an image», claiming that the Greek philosopher was referring to an initial process of knowledge through generalization (*IEI* [4], p. 119; *De Anima*, III, 7, 431 a-b). However, Simondon does not fully follow the empiricist thesis, since he does not believe that the process of induction is sufficient, especially when it is thought of as proceeding by accumulation:

> Inductive classes proceed, practically, from a germ, an absolute origin in experience; the whole class is given in the first experience, and the path of concrete knowledge resembles more an arborescent structure with more and more complex ramifications than a structure of summation (*IEI* [4], p. 120).

The generalization of memory-images is therefore not comparable to the induction thought of by empiricism since it proceeds by «asymmetrical ramifications» which, starting from a germ, generate the expansion of the reference class without eliminating the characteristics that distinguished it from others in perception (*IEI* [4], p. 123). Recalling *ILFI*'s argument, Simondon states that this proceeding by ramifications from a germ leads to a saturation that, through a leap, will lead to the symbol.[8] Saturation occurs when the different ramifications generated by successive perceptions become incompatible because they are opposite: «a metastable state that is the necessary condition for invention, that is, for a structural change restoring compatibility within a new system» (*IEI* [4], p. 124).

The memory-image can thus be compared to a «plurality of individuals» in which there is a single generic image or several generic images in relation to each other. It is precisely from this relationality within the image-remembering that the tendency towards the action of the symbol develops. There are, however, different ways in which the symbol tends towards action. Lacan's *Imago* is a type of image-symbol, since in the description in volume VIII of the *Encyclopédie française* it turns out to be that unconscious representation capable of organizing images (*IEI* [4], p. 127). In fact, the *Imago* is the equilibrium image that is constituted after the memory-image has become saturated. It succeeds, therefore, in bringing together different perceptions that were disparate within a new entity, a new system: the symbol. The latter is an entity that refers to a reality other than the *ego* and contains within it the different tensions that had manifested themselves within the image-remembering. If

[7] Simondon uses «mental images» as a synonym for memory-images.

[8] In *ILFI*, each individuation for Simondon originates from a germ that is structured within a saturated system and can only occur through a transduction that allows the transition from one order of magnitude (two incompatible systems) to another.

for Lacan the *Imago* is not a symbol because there are no three terms of the relation (a necessary requirement for Lacan), for Simondon, as an «organizer» that orders experience into a continuum, the Imago is already a symbol (*IEI* [4], p. 129). Also, in Jung's idea of the *archetype* Simondon finds the image-symbol and in this case the third term of the relation is present. According to Simondon, for the image-symbol to develop, however, it is not necessary for there to be a ternary or binary structure, but the individual is sufficient. For there to be image-symbol the only prerequisite is the «formalization of impressions». The only theory that, according to Simondon, contains an opening towards the connection between image and symbol is that of Piaget:

> In *La formation du symbole* chez l'enfant, connects the distinction between image and symbol to his distinction between functions of accommodation and functions of assimilation; yet there already exists in the image a constructive activity which entails that it is not simply an extension of perception [...] within a genetic perspective, does not exclude the possibility of forms of transition between image and symbol (*IEI* [4], pp. 130-131).

Beyond the symbol as mental reality analysed by Lacan, Jung and Piaget, Simondon considers it more useful for the purposes of the course on invention to analyse the objects that play the role of symbol. Invention is in fact a human activity that emerges from the symbolization of objects that are used. The first object to be used and then symbolized is the body. In this, Simondon seems to recall Merleau-Ponty and the importance he attached to the role of corporeality and carnality. In particular, for Simondon, the first use of the body is expressive imitation, which «awakens the animate object and brings it to life through evocation» (*IEI* [4], p. 131). Once the individual has learned to use his or her body as a symbol to evoke other objects, he or she begins to choose certain manipulable, removable and storable objects as new containers for symbols. This is the case, says Simondon, with clothing, or a cut strand of hair, or even a bit of water from a particular river (*IEI* [4], pp. 132–133), but even before that with organs outside the body that can be removed and adorned without harming the individual's life (*IEI* [4], p. 133). In cases where some individuals do not have many external organs that can be used as symbols, because they cannot be easily removed, they opt for a prosthesis (the sceptre, weapons, etc.). These objects become analogues of reality, key points through which the individual or the whole community can re-enter this reality. Referring to the third part of the *MEOT*, Simondon argues that the *analogon* function of objects can provide access to reality either through conscious knowledge or through a more primitive operation, that of magic. It is precisely this primitive magical operation that gives the *analogon*, especially the prosthetic, control over the symbol it represents, so that anyone who comes into possession of it is thought to be able to exploit its power. Now, these *analogon* or symbol objects are in an intermediate state between the organism and the object, because:

> It becomes a tensed and metastable system when the same prosthetic object, detached from any organism, is both the object which, carried by the other, addresses the subject, and which the subject tasks with transmitting the power of his own organism to the world (*IEI* [4], p. 135).

In this way, the initial memory image, which is asymmetrical because it contains several contrasting images, is transformed into a symmetrical system in which the same object-symbol acts both as a threat to the subject who uses it and as a weapon to threaten the other. Its metastability allows it to move to one side or the other, or to maintain both roles simultaneously. If, for example, it is the image-remembrance of a weapon in the hands of the subject who has the memory, or that generated by the weapon in the hands of someone else who threatens the subject, these are, for Simondon, simple images. The image-memory of the weapon becomes a weapon-symbol when a tension is created between these two images, «a symbolic weapon is all at once a threat against the subject and an object he can take hold of to threaten others» (*IEI* [4], p. 135).

Generalizing this example of the weapon-symbol and exploiting the evocative image of the weapon in profile, Simondon argues that the whole world of symbols is made up of objects «seen in profile», definable as a *pandemonium* between object and subject, between the living and the environment (*IEI* [4], p. 137). A non-primitive but naive magic uses this status of the object-symbol to try to make it concrete again, that is, to bring it back to life from only one side of the profile. This is the case of the *voult*, which is constructed from several object-symbols in order to reconstitute a being that can be real, and that magic seeks to re-enchant through spells and evocations. The term *voult* is taken from the original French version to avoid misunderstandings. In english, in fact, *volt* is the unit of measurement of electrical voltage and Simondon uses it in this meaning within the first part of *MEOT* [13]. But already in the second part of MEOT Simondon uses the term *voult* to refer to the *Encyclopaedia* as a magic formula capable of freeing techniques from the monopoly of workshops. In *IEI* the term he uses is instead *voult*, a now obsolete noun that can be translated in English as *face*. It is interesting to note that until the thirteenth century the term *voult* was used to indicate the face of God or some sacred figure, whereas from the fourteenth century it indicates witchcraft, i.e. a wax figure representing the person one wishes to injure or kill.[9]

Simondon argues that this is *naive magic* because he thinks he can reverse the cycle of the image and make an object-symbol an image again without the intervention of invention. In this sense, Simondon takes up Plato's critique of the phantasmal arts, i.e. those arts that, instead of producing good images of reality (εἰκών), produce illusory images (φάντασμα), and applies it to all those techniques that seek to reverse the cycle of images (*IEI* [4], p. 138; [9] 240a-c).

3.2.4 *Invention*

The fourth and final part of the course is dedicated to invention, the only way according to Simondon to close the cycle of the image in order to begin a new one. Invention is in fact triggered by a problematic situation that breaks the action

[9] Cfr. https://www.cnrtl.fr/definition/voult.

and requires a new compatibility between the organism and the environment and between the different directions taken by the action (*IEI* [4], p. 139). Invention allows, according to Simondon, a structuring return of the anticipation that was there at the beginning of the cycle:

> Sets up a certain kind of retroaction, a recurring input ("feedback") which goes from the regime of the completed result to the organization of the means and subsets within a mode of compatibility. [...] The root of the solution is a communication between two orders of magnitude, that of the result (the road reopened to all) and that of the problem-event (a barrier across the path of each one), whose data are altered [...] (*IEI* [4], p. 140).

The invention, therefore, on the one hand establishes a *feed-back* regime and at the same time brings into communication two orders of magnitude that had not previously found compatibility (transduction). Simondon gives the example of a group of travellers independent of each other who at a certain point find themselves in front of a boulder on the road they were travelling on. Each traveller can anticipate and imagine how they would continue walking if the boulder were to be moved, but this is not enough to allow the journey to continue. Only if all travellers combine individual anticipations and imaginations into a collective representation can they change the condition of the present and find a way to move the boulder. What happens is that invention is established, i.e. the compatibility between the collective and the individual outcome, but also between individual actions and the sum of the actions of others.

The different types of invention include deviation (finding a shortcut), the making of a tool and the association of several operators or individuals (collective or community). The case of deviation is present when one is faced with a situation in which one can choose different paths or behaviours to solve the problem: for example, when one has to learn the most valid strategy to get out of a maze (*IEI* [4], pp. 143–144; [6], pp. 205–218).[10]

The making or taking of objects in the environment as tools to solve a problem is a skill that, according to Simondon, is not the exclusive jurisdiction of human intelligence, as had been thought up to that point. But there is a difference between this inventive strategy present in non-human animal species and how it is exploited by the human species (*IEI* [4], p. 147). In fact, this idea that invention was a human prerogative was already present in antiquity, especially in Stoicism, which argued that each species had its own specific endowment to deal with the problems it encountered and that the human species was the only one to possess the inventive capacity as it was less endowed with other tools (organ-tools) (Le Moli, 2022, pp. 48–50; *Protagoras* 320c-322a). But this thesis has been carried on by scholars up to contemporary times, and in this regard Simondon cites several studies including those of Henri Piéron. The latter argued that one could not speak of invention in non-human animal species because every time an animal species uses tools outside its body to overcome an obstacle, it extends its instinct or uses a specific formula. The latter

[10] It is unclear whether Simondon was aware that in 1951 Claude Shannon had presented the *Theseus* mouse experiment at the *Macy Conference* in which it would appear that an artificial mouse had the inventive capacity for deviation that Simondon mentions in this passage.

would be a behaviour commonly found in a species, a stereotype that cannot be an invention. Simondon counters this argument by claiming that «the fact that a behaviour is encountered often within a species does not prove that it does not result from a multitude of concurring and parallel inventions» (*IEI* [4], p. 149). In fact, even in humans, inventions are usually non-uniform, i.e. not stereotyped, but under unfavourable conditions of poverty where possibilities are limited, stereotypies may not necessarily occur:

> In human life as in animal life, there is a constant necessity to confront the partial novelty of situations with an activity of organization of operative modes; the richer and more precise the mental image of the situation, the higher are the chances of discovering an adequate organization (*IEI* [4], p. 150).

In each case, human and non-human, the conditions for an invention to take place are the exploration, manipulation and organization of the territory. Only later can objects in the territory become tools for solving a problem. This happens because in the first phase of exploration, manipulation and organization of the territory, relationships were established between organisms and objects. These past relationships return as mental images in the organism faced with the problem and an interaction is generated between past experiences and the current objective (*IEI* [4], pp. 151–152). Simondon argues that the field of past experiences, although more numerous, has a smaller gradient (a tendency or slope) than the gradient of the goal, which, on the other hand, manages to attract the smaller gradient by organizing it. This condition is referred to by Simondon as *amplification*, in the manner of that found in scientific field theory. A smaller field but with a stronger gradient manages to amplify the larger field but with a weaker (almost neutral) gradient (*IEI* [4], pp. 151–152). This implies that images acquired with disinterest on the part of the subject can be a more solid basis for structuring an invention because their polarity, their gradient, is weaker, tending towards neutrality, and therefore more easily organized into a new system.[11]

Simondon argues that there is a purely human sphere of invention, and it is that which concerns «objective metrological formalization», i.e. the transition from technique to science (*IEI* [4], pp. 154–155). Initially, this transition was prepared by using not only objects but other living beings as instruments to achieve an objective. When man began to also use other human beings as aids (slavery), subjugation required a further step. To make one subject the slave of another and thus in a position to submit to the orders of a subject to achieve a goal, Simondon considers it necessary to communicate the entire project when trying to achieve it. A further step was taken with the help of machines that received orders all at once and for which man had to find a way to communicate through rules and data. Finally, the need for formalization became even greater when learning was no longer manual, as in the workshops, but when it moved into schools where formulas were needed that could be expressed with numbers and figures that could be understood by everyone. It was that universalization of knowledge that came about with Diderot and D'Alembert's *Encyclopédie* (*IEI*

[11] This is the saturation condition described at the end of the third part of the course and allows the transition to invention.

[4], p. 154). The same formalizations that had initially served to support an asymmetrical system (master–slave) began to structure themselves to become a *universal symbolism* that allowed for different operations and the construction of an abstract world made up of «representations of objects and formulas of relationships» (*IEI* [4], p. 156).

Alongside objective formalization, Simondon recognizes a subjective formalization that concerns the affective-emotional aspects of the relationship between the subject and the world. These relations cannot be expressed through operations, but only through individual or collective actions that communicate feelings, motivations and emotional drives. These actions are the rituals and norms of leisure time that alternate with the operations of work time. Thus, a dichotomy is generated through religion (*IEI* [4], p. 157).[12]

The various inventions of formalizations serve to ensure control, governance but also the modulation of larger realities by smaller ones. Man invents formalizations to guarantee himself a paradigm for understanding larger orders of magnitude that he would otherwise be unable to comprehend. Simondon compares this act of invention to the growth of living organisms: both are characterized by organizations that amplify the resources received from the environment (*IEI* [4], p. 162). In the image cycle, once the memory-images have accumulated generating a saturated system, there is a reserve of content in a latent state but rich in potential for transformation. This reserve is structured, organized to give space to the potential contained therein.

The process of invention, however, can only be said to be truly formalized when it produces an object that can be separated from its inventor by becoming communal. In this sense, although Simondon does not want to deny the presence of culture in non-human animal species, it can be said that this level of invention is peculiar to humans since ours appears to be the only species in possession of the necessary means of transmission. This means that non-human animal species can invent objects, but are not able to make them independent and, therefore, available for uses other than those for which they were invented by other beings of the species and subsequently (*IEI* [4], pp. 163–164). This process of inventing objects and transmitting their use allows, according to Simondon, progress as a network of inventions that build on each other. If at the end of the 1950s during one of his lectures in Poitiers he had strongly criticized the concept of progress as mere improvement [7, pp. 83–99], here he considers the concept of progress as alternating between regression and advancement to be well established, emphasizing the necessity of the interaction between culture and the production of objects for it to take place:

> The created object is precisely an element of organized reality that is detachable because it was produced according to a code belonging to a culture enabling it to be used far from the time and place of its creation (*IEI* [4], pp. 163-164).

Simondon, therefore, emphasizes the need to reintegrate the production of objects into culture so that we can understand the universality and timelessness of the invented

[12] In these passages, one cannot fail to notice an assonance with the description of the various phases and dichotomous divisions analyzed in the last part of *MEOT*.

object (virtuality) and, consequently, the possibility of reusing and reincorporating objects into new inventions (progress). Simondon in this text criticizes the concept of progress proposed by both Marx, Bergson and Weber because he does not find in it a real attention to the constructive side of the crisis. Every phase of progress is made up of a moment of regression in the absence of which one cannot proceed towards a new organization. Progress, if thought of correctly, constructs set, de-differentiates them, disorganizes them and then reuses some of the elements present with a different function in the new organization.

In fact, this hardly occurs because the invented object is thought of as a prosthesis of its inventor, impossible to detach. Simondon takes up here the example of the automobile already analysed in *MEOT*, arguing that from the technical object created «in purity» to the one that is used on the streets every day there are a series of superficial layers that create a semantic contact with man. It is precisely these merely aesthetic layers of manifestation and the intermediate layers (which Simondon calls more *semantic*) (*IEI* [4], p. 165) that hinder the virtuality of the invented technical object. It is also true, however, that these aesthetic and semantic layers that are superimposed on the «wild» technical object provoke the regression that serves as the impetus for the search for new inventions. For Simondon, in fact, invention is not produced to achieve an objective, but to solve a problem that arises. Therefore, it must be accepted that it is not possible to know in advance the effects of an invention if it is really an invention and not a mere addition of aesthetic or semantic layers.

Recall Simondon's criticism in *MEOT* of car companies that by only changing the aesthetics of the object try to sell it as a new invention. The comment that Jean Baudrillard inserts in *The System of Objects* after quoting a passage from Simondon's *MEOT* also moves in this direction. Baudrillard insists on the need to find a different point of view on the technical objects that are part of everyday life and, as such, can be influenced by fashions:

> We may dream of arriving at an exhaustive description of technemes and their semantic relations that would cover the entire world of real objects, but this must inevitably remain just that—a dream [...] this impulse immediately runs into the directly experienced psychological and sociological reality of objects, a reality which, over and above objects' perceptible materiality, constitutes such a significant body of constraints that the integrity of the technological system is continually being modified and disturbed by it. It is this disturbance, and the way the rationality of objects comes to grips with the irrationality of needs, and the way this contradiction gives rise to a system of meanings that seeks to resolve it—it is these things that we are concerned with here, not technological models, even if the essential truth of these models provides the ground from which our direct experience of objects is continually emerging [8, p. 7–8].

In true invention, says Simondon, there is «a leap» that has an amplifying power not limited to mere purpose and momentary adaptation (*IEI* 4, p. 171). According to Simondon, therefore, every time an invented object appears (thus detached from its inventor), a process like that which occurs in Lamarck's theory of evolution is triggered: one incorporates this object into reality and the random effects of its invention become regular functions (*IEI* [4], p. 173). The invention initiates a process that overcomes the problem from which the invention itself started and goes beyond the idea of a direct and strict purpose.[13] This going beyond direct purpose is defined by Simondon as *functional surplus value* and occurs from the natural realities incorporated in the invented technical object: «nature is re-created as necessitating formalization and concretization within the universe of technics» (*IEI* [4], p. 174).

3.3 Conclusion: A Conceptual Laboratory for a Theory with *Psyche*

An analysis of the 1965–1966 course *Imagination et invention* allows us to grasp the full richness of Simondon's mature thinking on the psyche and, more generally, to appreciate the centrality of psychology within his philosophical project. Far from considering imagination as an ancillary or secondary faculty, Simondon recognises its structuring role in the process of psychic and collective individuation. For Simondon, imagination is not a simple faculty of representation, but a dynamic and transindividual function that lies between the real and the possible, between the information received and the creation of meaning, between affective experience and technical production.

In the course, Simondon develops a conception of imagination as a metastable and stratified process that traverses different levels: the emotional, the symbolic, and the technical-inventive. In this sense, imagination is an *ontogenetic operator*: a vector that, through states of tension and phases of instability, allows the passage from one level of being to another. It is, to all intents and purposes, a *transformative energy* that acts in subjectivity as in the world, and which makes the concretisation of invention possible.

Invention, after all, is the place where imagination fully unfolds its function: it represents the subject's active response to an open problem, a *form of mediation* between the demands of the environment and the internal structures of the individual. Invention is never purely technical, but always involves an affective movement, a creative gesture, a restructuring of the relationships between oneself and the world. In this sense, the link between imagination and invention reactivates the profound connection between psychology and technology that runs through Simondon's entire work, showing how psychic life is not a separate realm, but a place of articulation between the dimensions of reality.

The originality of Simondon's proposal, as emerges from this course, therefore consists in overcoming both mechanistic conceptions of imagination, reduced to the manipulation of images, and romantic visions that reify it as a spontaneous source of genius. For Simondon, imagination is an *intermediate, processual, collective and relational function*, and it is precisely for this reason that it can constitute the *cornerstone* of a new genetic psychology. In it, subjectivity is understood as a field of

[13] In this passage, although Simondon does not say so, it seems possible to recall the idea of Darwinian natural selection whose purpose is not direct since it acts following certain laws, but without knowing in advance the effects of its action.

tensions and transformations, in which affectivity, perception, the symbolic and the technical are constitutively intertwined.

Imagination et invention thus shows how Simondon is undertaking a veritable epistemological refoundation of psychology, proposing to reinterpret it on the basis of processual, transindividual and ontogenetic categories. This is a highly topical proposal, which opens up a more complex understanding of human creativity, the role of technology in subjectivity and the ways in which new forms of meaning are produced. Far from being a simple "minor course", this text represents a *conceptual laboratory* in which Simondon tests some of his most fruitful insights, projecting psychology towards a new horizon, capable of thinking together about affectivity, symbol, technology and invention.

In this sense, imagination reveals itself as one of the privileged places for understanding genetic psychology in action: not so much a theory *of* the psyche, but a theory *with* the psyche, understood as a point of passage and a field of resonance between the different regimes of being. The analysis of this course therefore invites us to continue research into the role of the psyche in Simondonian philosophy, starting from a notion of imagination that, far from being static, is configured as the beating heart of individuation and transformation.

References

1. Simondon G (2014) Sur la technique, 1953–1983. Puf
2. Simondon G (2004) Deux leçons sur l'animal et l'homme. Ellipses
3. Simondon G (2015) Sur la psychologie, 1956–1967. Puf
4. Simondon G (2022) Imagination and invention. University of Minnesota Press [IEI]
5. Simondon G (2018) La résolution des problèmes. Puf
6. Sunseri F (2022) Algoritmi viventi tra Shannon e Simondon. Lo Sguardo 34(1):205–218
7. Simondon G (2016) Sur la philosophie, 1950–1980. Puf
8. Baudrillard J (1996) The system of objects (trans: Benedict J). Verso (Original work published 1968)
9. Plato (1996) Sophist (trans: Brann E, Kalkavage P, Salem E). Focus Publishing
10. Simondon G (1964–65) L'instinct. Bull de Psychol 18(235):75–103
11. Simondon G (2006) Cours sur la perception (1964–1965). La Transparence
12. Simondon G (2010) Communication et information: cours et conférences. La Transparence
13. Simondon G (2017) On the mode of existence of technical objects. Univocal Publishing [MEOT]
14. Simondon G (2020) Individuation in light of notions of form and information. University of Minnesota Press [ILFI]

Chapter 4
The Reception of Simondonian Psychology

4.1 Introduction: The First Debates in France

Since the defence of the theses in 1958, the reception of Simondonian thought has followed different paths due to the timing of the publication of his works. While the complementary thesis on technical objects (*MEOT* [47]) has been quite successful since 1958 due to its innovative look at the question of technique, the main thesis about individuation (*ILFI* [48]) has been fragmented into late, disconnected publications (*IGP* [46], 1964; *IPC* [45], 1989; *ILFI* [48], 2005) and not always placed in relation to the complementary.

To think that Simondon produced two doctoral theses during the 1950s that were not deeply related was certainly the greatest sin of his readers. But it must be said that Simondon himself never tried to make the connection between the two works any clearer. One only has to reread his interview with Jean Le Moyne in 1968 and the reference to «hasards universitaires» [1, pp. 405–445] in response to the question as to why he had dealt with different themes in the two theses. After the relatively isolated cases represented by Laloy and Van Caneghem, the connection between the two works will be explicitly placed at the centre of the interpretative project developed by Jean-Hugues Barthélémy, who has always insisted on emphasising the strong encyclopaedic spirit that guides Simondon [2].

The absence of a systematic in-depth study of the connection between Simondon's works in the early years of his reception and the underestimation of the encyclopaedic instance that animates his *corpus* can also be considered the causes of the lack of attention paid to the intense work on the psychological side that the author conducted from the very first years of his teaching and scientific magisterium.

Even though Simondon taught general psychology for many more years than philosophy, his commitment to psychological research received little critical attention. Apart from the space that the Bulletin de psychologie granted him in the 1960s to publish his lectures, no specific studies addressed this aspect of his work until 1989,

F. Sunseri, *Psychology and Technology in Gilbert Simondon*, UNIPA Springer Series,
https://doi.org/10.1007/978-3-032-25740-6_4

the year of his death. It was only then that the texts he had published in the Bulletin—in which he elaborated his theory of psychic individuation, expanding on the work undertaken for his 1958 thesis (*ILFI* [48], pp. 257–326]—were cited for the first time. Interest in these lecture reports is developed in the aforementioned article written in homage to the figure of Simondon and published in the same Bulletin by Denise [3]. In this study, Simondon's pupil retraces the author's life and works, emphasising the difficulty of reception, particularly with respect to psychology studies. The article opens with an initial acknowledgement by Van Caneghem of Simondon's encyclopaedic vocation and how this was also to some extent the cause of the partial oblivion into which his work fell:

> [...] Simondon was a true Philosopher, that is to say a man who drew his defences against universal anxiety not only from the effort to unify knowledge, but also from its transmission, in the midst of students and colleagues in good company, as in the time of the "Enlightenment" [3, p. 817]

It is thus reiterated how Simondon's effort turns out to be primarily that of unifying knowledge in order to then transmit it to all those who would follow him, as Van Caneghem argues, just as the knowledge brought together by the *Encyclopédie* of the Enlightenment was open to all. The problem traced by Simondon is that of the relationship «between such and such a technical object at a given time [...] and the individual reality of such and such a living being, also unique and yet very much situated in a culture at a given time» [3, p. 818]. Van Caneghem traces the originality of Simondon's analysis in the particular look he casts on this relationship. On the one hand, man's representation of himself in technical objects (recall the idea of the robot criticised by Simondon in the opening of *MEOT* [47], pp. 16–17) is the starting point for investigating the way human beings think about the structuring of their psychosomatic and psychosocial functions; on the other hand, the study of technical objects, their genesis and evolution, allows Simondon to open up a profound reflection on the amount of «clarté et opacité» that the technical object has within itself and to show how it is nothing more than the transposition of something that belongs to the human inventor of the objects themselves.

These two original traits of Simondon's thought intertwine from the need to rethink the process of individuation on its different levels (physical, living, psychic and collective). For Simondon, retracing the ontogenetic path from which what is usually called the *individual* emerges means first of all demolishing piece by piece the substantialist and atomist conceptions according to which there is an inseparable unity (the substance or atom) at the bottom (*ILFI* [48], pp. 21–133). One of the best-known formulas in the introduction to *ILFI* is that it is necessary «to know the individual through individuation rather than individuation starting from the individual» (*ILFI* [48], p. 3). By this Simondon wants to emphasise the importance of starting from the process that allows for individuation without presupposing the existence of an individuated individual as the beginning of the process itself. This problem, analysed in the first chapters of the first section of *ILFI*, conceals another issue that Van Caneghem intercepts and concerns Simondon's search for a unity of the human sciences. There were several theories or models that could explain the different functioning of the living being (Darwin's organ-mechanist model, Janet's organ-dynamic model, or even the psychoanalytic and phenomenological models) [3, p. 821], but none of them could explain the entire ontogenetic process, the functioning of which did not differ, according to Simondon, from domain to domain. Moreover, none of these models could predict the interaction between two systems, for example the behaviour of a given human in a specific situation. This inexplicability depends on the deep connection between the systems at play, which cannot be studied separately from one another and then attempt to explain one through the other. Simondon's is thus, once again, a vocation for Renaissance wisdom or the spirit of encyclopaedism, in which there was not the scientist or the philosopher, but the sage or the scholar who approached the systems to be questioned without specific limitations. In *MEOT*, Simondon defines the encyclopaedic spirit as «movement [*mouvement de fond*] expressing the need for attaining a state of freedom and adulthood» (p. 112) and considers that this movement has been present in history three times. The first is that of the Renaissance during which, in order to move from a status of minority (the child) to one of majority (the adult), knowledge was expanded through the search for the true text of the Bible and the Greek texts of philosophical works. The second time this movement of liberation appears is during the Enlightenment when, thanks to the *Encyclopaedia*, techniques were liberated through the use of visual language, the diagrams (*MEOT* [47], pp. 112–115).

Van Caneghem compares the effort of the individual human sciences to find models capable of explaining the processes of individuation to that «résonance intime» that the subject experiences when he feels divided and tries to find resolution [3, p. 821] within himself, and which Simondon identifies with anxiety (*ILFI* [48], pp. 282–284).

Van Caneghem proceeds in the analysis of the entire Simondonian theoretical endeavour by traversing the works published up to that time and concentrates on the second part of *ILFI*, the one dedicated to psychic and collective individuation published for the first time that very year. What emerges from this analysis is the Simondonian proposal to start again from perception by setting consciousness aside. In this sense, the influence of Maurice Merleau-Ponty's lectures on the young Simondon and the decisive move away from a classical Husserlian phenomenological perspective is clear. It therefore appears necessary to start again from perception as a problematic moment that in the living provokes the drive towards the structuring of psychic individuation. This is not, however, a structuring without genesis [3, p. 831] as proposed by *Gestaltpsychologie*, but an attempt to phase two metastable systems, the one formed by forms and background and the one constituted by the individual and his associated environment. In this sense, Van Caneghem sees in Simondon a distancing from Freudian psychoanalysis in favour of Jungian psychoanalysis, since there would be no idea of an unconscious that is formed from the conscious. For Simondon, the unconscious remains in fact that which of the pre-individual continues to resonate and problematise in the living, while the transducing centre between the unconscious and the conscious becomes the subconscious, i.e. the locus of affectivity. Following Jung, says Van Caneghem, «les centres hypothalamo-hypophysaires», i.e.

the place where the subconscious would be located, would not only allow the connection of the conscious perception with the unconscious part, but would also open up the possibility of the process of individuation and collective relationship [3, p. 831]. It is precisely because of this openness that Van Caneghem identifies an indissoluble link between psychic and collective individuation in Simondonian thought. The process of psychic individuation would remain incomplete without communication, that moment of reflection (in the sense of Merleau-Ponty's looking in the mirror) that each individual makes towards the other. This moment of contact between the two phases of individuation is, in fact, a crisis, a moment of splitting that can lead to the metastable resolution of the collective, but can also generate «narcissistic fascination, by mirror-images too quickly fulfilling an old desire for eternity, i.e. self-sufficiency and death» [3, p. 831].

A limitation of this—in many ways pioneering—critical contribution is perhaps that of not fully valorising the difference between affectivity and emotionality, which, instead, is particularly central in Simondon's work, particularly in the very sections of *ILFI* examined here. Evidence of this is a failure to theorise the specific role of affectivity, something that will instead be the focus of later research [4–10]. This first study is nevertheless of fundamental importance in that, starting from the focus of psychology, it succeeds in grasping another fundamental aspect of Simondonian thought, destined to be taken up by Bernard Stiegler [11–13, 49] among others: the role of *temporality* in the process of individuation in general and in particular its central function for those levels that would seem to belong exclusively to human individuals: the psychic and the collective. In fact, Simondon maintains that the process of individuation is characterised precisely by the plurality of times between which that displacement is generated from which the individual is a phase «the result of a series of integrated phase shifts» [3, p. 832]. It is only in the collective, through trans-individuation, that this plurality of times can be «looked at», i.e. loaded with meanings through the process of symbolisation and allow the individual to recognise himself as a phase of an ontogenetic process that starts from afar and leads towards further dimensions.

Van Caneghem's work is therefore important as the first step towards the discovery of Simondonian psychology, which, as can be understood from the encyclopaedic spirit that accompanies it, cannot be detached either from the general theory of individuation or from the view on technique. In closing the article Van Caneghem recalls how Simondon was, together with [14], one of the few who believed in the possibility of an articulation of psychological models to arrive at a truth about man [3, p. 836]. This can be proven by the use (so present in the last part of *MEOT* [47]) of the concepts of network or reticulation (*MEOT* [47], pp. 29 ff.; [15][1] and by the importance Simondon attached to relationality and communication as the only possible way to a conscious knowledge of the relationship between man and the world.

[1] Simondon introduces the terms network and reticulation to speak of technical ensembles that go beyond the scale of human experience, but also of the condition for a technical object to fully unfold.

Shortly after Simondon's death and the first publication of the second part of *ILFI*, the idea appeared for the first time in Van Caneghem's perspective that the readings of Simondon's work up to that point were the initial profiling of a research that, in order to even correspond to the encyclopaedic, systematic and *allagmatic* intentions of the author, had to include a specifically psychological study of individuation processes (conducted in *ILFI* [48]) and the analyses on the technical object (contained in *MEOT* [47]). We must therefore wait and hope for the publication of the other works in order to further explore the issues Simondon included in the two theses. We are still a long way from the idea of a Simondonian psychology as a possible «key» to understanding the relationship between the two theses and the perspectives of thought outlined in them, but the direction is open.

A further space explicitly dedicated to the analysis of psychic individuation (and thus to the specifics of psychology) prior to the publication of the other Simondonian works on the subject is that contained in the monograph that Muriel Combes dedicated to Simondon in 1999. This is a reconstructive work on Simondon's thought contained in *IPC* with particular attention to the concept of the transindividual [16]. What stands out in the second chapter devoted to the transindividual relation is the attention with which Combes emphasises the Simondonian position regarding the incomplete identification of psychic individuation as a true individuation in its own right:

> To be rigorous, we must say that "*there is not, strictly speaking, a psychic individuation*, but an individualisation of the living which gives rise to the somatic and the psychic" (*IPC*, p. 134; emphasis added). Psychic individuation is a perpetuated vital individuation [16, p. 30].

In fact, psychic and collective individuation do not lead to new individuals, but are the extension of the living individuation that led to metastable individuals who, in order to exist, must be able to continue to individualise themselves by solving the problems of their surroundings (their associated environment). As Simondon points out, there is no psychic individual detached from the living: there is no soul without a body. This conception allows Simondon to review the relationship between man and animal no longer as, traditionally, a difference in kind, but only one of degree. Combes emphasises how, in the Simondonian perspective, animals also have a certain degree of extension of living individuation, individualisation, which allows them to organise their reaction according to the space they have around them. What marks the transition from animal to human is therefore not the presence of individualisation, i.e. the possibility of structuring or organising at a more complex level, but the animal's inability to represent the space around it as detached from the space itself. In other words, the animal is unable to symbolise, i.e. to detach meaning from the object it is given and then attach it to something else, or it seems perpetually to feel without being able to rise to the level of the representation of the object separate from contact with the object itself.

Combes highlights the fundamental function that Simondon assigns to the affective-emotional layer as the domain of intensities in which individuality resides. Combes notes a connection between Simondon's psycho-social thinking in these pages of *IPC* with Spinoza's conception of the subject of ethics, i.e. the one who is able to be the cause of affects in others by increasing or decreasing their power to act, but also to be affected by the affects of others that generate the same change in degree of power of action. Combes points out that in Spinoza, as in Simondon, consciousness does not play a pre-eminent role because it is not stable and autonomous, but changes according to the subject's affects (internal or received from the other).

4.2 Bernard Stiegler and Epiphylogenesis

One of the first scholars of Gilbert Simondon's thought to highlight aspects related to psychology was Bernard [17]. In his 1998 article, Stiegler addresses in depth the question of individuation in Simondon, showing how technology, as a fundamental dimension of human existence, is not a mere object of study but the very principle through which both the psychic individual and the collective individual are constituted. This implies a significant reinterpretation of *psychologie simondonienne*, not as a simple psychology of the individual subject, but as a dynamic philosophy of individuation that explicitly intertwines the technical sphere, the psychic dimension, and the social dimension. It is by now well known that Stiegler's analyses of Simondon's thought are not "pure", but are part of a work of interweaving—pursued by Stiegler from his very earliest writings—among authors who are often very distant from one another.

The 1998 article opens with a reminder of the non-static nature of the technical object in Simondon's philosophy: «Simondon décrit l'objet technique comme tendant à l'organisation, comme s'individuan» [17, p. 241]. This definition of the technical object as something that individuates, an entity in process, is fundamental for Stiegler in order to affirm that the process of individuation does not concern only living individuals. This process is made possible through transduction, which, from the "preindividual horizon" tends toward individuation without ever fully completing it.

Stiegler acknowledges that the same type of operation is also applied by Simondon to psychic and collective individuation, but he believes that a fundamental step is missing that would connect the psycho-social dimension with the technical one: «Simondon n'a pas articulé ces deux aspects de son travail» [17, p. 241]. According to Stiegler, this lack of articulation represents a significant gap, since technology is what effectively mediates between the individual subject and the collective dimension, insofar as it is at once «trace and heritage», that is, external memory and a repository of accumulated potentialities. We can now say that Stiegler's position on this occasion perhaps suffered from an incomplete reading of Simondon's work, since the courses from the 1960s clearly show precisely this connection. It is also true that the *Cours* were published several years later and that, therefore, Stiegler could not easily have had access to this material in 1998. It is in light of this Simondonian lack that Stiegler employs the concept of epiphylogenesis—already introduced in his early articles of 1992 [18, 19]—to designate the way in which technology mediates psychological and collective individuation. By this he suggests that technology is not a secondary phenomenon with respect to human life, but the *conditio sine qua non* for the formation and continuity of psychic and collective structures.

Stiegler further argues that if technical objects evolve through transduction by passing through phases of concretization, then the human psyche itself may also be considered as the product of the same processes, rather than as a preformed entity merely receptive to external stimuli [17, pp. 243–244]. The human psyche is therefore not a stable core from which actions and relations derive, but is itself an ongoing process, constituted by dynamics of transduction that incessantly connect it to a broader context of preindividual and collective forces.

For Stiegler, the individual is a metastable equilibrium, always in *déphasage* with itself and with its environment: «l'individu est un équilibre métastable [...] qui se maintient dans la tension de l'inachèvement qui commande toute individuation» [17, p. 246]. Stiegler reinterprets Simondonian psychology as a psychology of becoming, that is, the study of a process in which temporality plays a fundamental role. The psyche and the individual are never stable points of arrival, but continuous negotiations among forces and tensions that are organized over time through the mediation of the technical world. The crucial point of Stiegler's reading lies precisely in this *metaxu* role played by technical objects:

> La question se pose alors de spécifier les conditions de *conservation* de l'individualité plus vaste que les individus psychiques: à travers eux, mais plus vaste qu'eux, et ici, *hors* d'eux. Cet élément plus vaste est pour l'individu moins vaste un déjà-là prothétique, c'est-à-dire: instancié dans les objets techniques constitutifs d'un monde [...] supports de tradition et de savoir, d'un passé 'historial' [17, p. 249].

Technical objects thus play the role of repositories of external and symbolic memory, and this allows psychic individuals to enter into the collective by constituting themselves as subjects rather than merely as individuals. In this sense, Stiegler reads Simondon with his own concept of epiphylogenesis already in mind, and maintains that the collective should not be thought of as an individuation that follows the psychic one, but as an «emboîtement»[2] of simultaneous individuations [17, p. 249]. The interlocking of psychic and collective individuation goes beyond the "merely" living individual and retains that memory externalized in technical objects which the single living individual would lose at death. This type of psycho-collective memory then provides the basis for selection in the reproduction of the species: «l'épigenèse exerce un puissant effet en retour sur la reproduction de l'espèce» [17, p. 250]. Technology, psychology, and collectivity thus appear to be closely linked in a transductive relationship that is no longer binary but ternary [17, p. 251]. In Stiegler's reading, technology becomes the condition of possibility of collective psychological experience and historical memory: it innervates our understanding of the world and of ourselves through the externalized traces that precede us and that constitute us as subjects. Stiegler acknowledges Simondon's merit in having attempted to look in this direction through the concept of the transindividual, but considers this attempt still superficial. In fact, Simondon maintains that the transindividual allows us to see the reciprocity of psychic and collective individuation, taking for granted that

[2] *Emboîtement* can be translated as joint or connection.

one understands that technology lies behind transindividuality. Indeed, by reading *ILFI* and *MEOT* together—as if they were a single work—one could easily grasp Simondon's intention, yet the fact remains that without access to the courses from the 1960s everything remains open to interpretation.

Thus, Stiegler emphasizes the need to pass through epiphylogenesis in order to grasp the reciprocity of the psychic and the collective [17, p. 251]. Moreover, in order to underline the profound connection between technology and the psycho-collective, Stiegler argues that, for the process of concretization of technical objects to occur, there must first be an anticipation on the part of human beings. At this point, Stiegler ventures into a critique of Simondon, whom he accuses of having considered technicity as external to temporality—thereby showing himself to remain too closely tied to Bergsonian terminology [12]. In truth, I believe that Simondon considers technicity not in terms of the essence of technology—a reference that Stiegler almost certainly takes from Heidegger, whom he discusses in the subsequent pages of the article—but rather as a "preindividual" ground. Perhaps Stiegler himself had already intuited this when he wrote that «tout se passe d'ailleurs comme si les analyses de Simondon le montraient sans que lui-même ne le voie» [17, pp. 251–252].

4.3 Pascal Chabot and *l'œuvre Au Noir* Di Simondon

The condition desired by Van Caneghem began to be realised from the early 2000s onwards, when not only was Simondon's oeuvre brought together (*ILFI* [48] is from 2005) and analysed in its entirety, but the articles, courses, lectures and interviews that the philosopher had written and held after 1958 and up until his death began to be published. The publication of these texts in collections sheds a first important light on Simondon's relationship with psychology and the latter's role in the construction of the overall systematic edifice. Surrounding these publications are, in fact, the works of scholars who will increasingly emphasise Simondonian psychology not only as a specific feature of his main doctoral thesis but as an integral part of his philosophical system, what Barthélémy calls «encyclopédisme génétique» [2]. Still absent, in the contributions of these years, is the idea of a centrality of psychological reflection to understand the tightness and articulation of the author's overall construction of thought.

The first monograph dedicated to Simondon in which an entire section on the psychological question is included is Pascal Chabot's *La philosophie de Simondon* from 2003, in which the Simondonian philosophy of technique is declined according to the directives of the new digital objects [20, 21].

In the 2003 text, in the first and second parts, Chabot analyses the thought on technique and that on individuation following the example of his predecessors. The third part, entitled *Les ponts*, is dedicated to the psychological question as a possible point of contact between the two theses: «the bridge linking them is not drawn» [22, p. 105]. Chabot's attempt is to find such a contact, starting by bringing into dialogue the psychologist Gustav Jung's concept of individuation with Simondon's, which he calls 'depth psychology' because it delves beyond that dark zone already outlined in the first pages of *ILFI*: the zone between consciousness and unconscious in which the drive towards individualisation and then towards the collective develops. In the first phase of his estrangement from his master Freud, Jung considers individualisation a «chemin de crête», a ridge that not everyone is able to cross because not everyone is able to reunite the ego with the self, the conscious with the unconscious [22, p. 107]. In this perspective, Chabot believes that Jung speaks of individuation as a guilt that the subject must atone for in order to gain access to society. The only way to do this is to find a balance between the conscious and the unconscious. This balance must be indestructible, i.e. generate an 'in-dividual' in the atomistic sense [22, p. 109].

It is only in the 1930s that individuation begins to assume different characters in Jungian thought and, as Chabot points out, closer to Simondon's perspective. The process of individuation begins to be characterised in differentiations as stages through which the individual loses his or her fixed points and then in stages of the emergence of the self, no longer as the centre of the egocentric individual but as the starting point for the individual to find his or her place within the world. In this sense the individual of the mature Jung is never unitary, but always dual: on the one hand consciousness, on the other the unconscious. An individual, Jung's, who is nevertheless bound to a «the feeling of being part of something that wasn't me» [23, p. 87].

Chabot recalls a passage from *IPC* in which Simondon refers to Jung and that *œuvre au noir* of alchemical derivation that inspired his mature conception of individuation: that phase of return to a state of birth, rich in potential, which can open up new vital structuring. Chabot uses these references to compare Simondon's work on the concept of individuation to the *Grande Opera* of alchemy during which the individual is seen as a diamond in the rough destined to mature by freeing the spirit. In this sense, Simondon analyses the classical concepts to purge them of interpretations that distort their origin. This seems to have happened in particular with the concepts of the individual and the psyche.

Chabot is the first to speak of a «livre de Simondon sur la psychologie» with reference to *IPC* [22, p. 111] and argues that the real *coup de théâtre* occurs in the difference he found between individuation and individualisation. It is in this difference that the break with Jungian thought occurs. If for Jung, individualisation lies at the centre of life and follows its course through the different stages that bring the individual closer to society, for Simondon, (psychic, in this case) individualisation is a problem to be solved that arises from the moment the living individual feels double, split into a part that binds him to his corporality (perception of the external world) and a disparate part that detaches meanings from the perceptual world and continues to make them live within. It is the process of disparation between perception and symbolisation in which, for Simondon, lies the only real gap between other animals and man. This *disparation* cannot be explained, according to Simondon, either through transcendence or immanence since neither perspective considers that, prior to individuation, there can be that complete reality defined by Simondon as the dimension of the *preindividual*.

The final move that allows Chabot to coherently complete the *pont* between the two theses and «souls» of Simondonian thought is to connect *IPC* with the critique of the work present in *MEOT* by arguing that Simondon reactivates an archetype of technical intuition [22, p. 120], a magical trait of technique that has, since time immemorial, fascinated man. Simondon is at a moment in history when consumerism was in its infancy and out of control. It is in this consumption, in the mindless work that is found in the vicious circle of producer–consumer, that Simondon sees the risk of loss of value of quality in favour of quantity. Says Chabot, «the problem is primarily psychological» [22, p. 121], as it derives from the division between subject and object made by Western culture. Immersed in the triumph of scientistic rationality, Simondon proposes instead to return to look at ontogenesis to find the co-presence of the rational and the irrational, the continuous and the discontinuous, but also the automatic and the dis-automatic. The ontogeny traced by Simondon tells of intuitive, qualitative instants that give rise to those individuals in whom one always attempts to find only rationality (human beings, but also technical objects).

4.4 Jean-Pierre Esquenazi and Affectivity as a Hinge of Individuations

Thanks to the analysis of the section dedicated to the psychic identification of *ILFI* and the magical-symbolic state present in *MEOT*, the centrality of the affective dimension in Simondon's philosophy has been highlighted. This centrality forms the starting point for the work of the French sociologist, Jean-Pierre Esquenazi.

Referring to the partial *IPC* read through Deleuze's theses, the author makes a lucid analysis of the intervention of affect and emotion in the dynamics of individuation [6]. Esquenazi's intention is to reconsider the value of emotion with respect to an excessive exaltation of rationality in the relationship between the collective and the mass media [6, 24–28].

This work fits in perfectly with the general intention of Simondon's work of the foundation of a new humanism based on the reconsideration of scientistic rationality as a bad derivative of the faith in reason of the Enlightenment, which was instead aimed at freeing techniques from the prerogative of workshops [29]:

> It seems to me necessary to reject an overly unequal division between rationality and emotion, which would only enhance the former and depreciate the latter. Whether we like it or not, our reactions are wrapped up in emotion [6, p. 47].

Esquenazi would like to rethink the emotion that is provoked by topical images transmitted by the mass media not in terms of quantitative calculation, but with respect to the possibility of generating collectivity only from the prolongation of emotion itself. In the article's introduction, he associates emotion with a key term for Simondon: «an emotion could be [...] a phase» [6], 47). The sense Esquenazi seems to give to the concept of phase differs from Simondon's, but the conclusion they reach seems to find several points of contact.

Esquenazi's starting point is Gilles Deleuze interpreter of Simondon, who argues in *Différence et Répétition* that in order to set a reflection in motion, it is not enough to have an object standing in front of a subject, but something is needed that emanates from the object and from which the subject cannot escape [30, p. 182]. Deleuze speaks of «tonalités affectives» that pose themselves as a problem to the individual who encounters the object and that lead him to seek a resolution. Reflection, thinking, therefore, is subject to a feeling, to something that bursts forth in a violent manner without the individual being able to escape it.

Starting from the example of people's reaction to the images of the terrorist attack on 11 September, Esquenazi proposes the Simondonian theory of emotion as a vehicle to demonstrate how the disruptive affect of the images broadcast by the mass media in the hours following the attack did not immediately have a follow-up because an emotion, i.e. a collectivity that gave meaning to the event, was not generated. This means, according to Esquenazi, that the affection generated initially cannot suffice to open up reflection, but «it must be preserved, but in a different form» [6, p. 50]. This different form that affection must preserve is generated from a process that Esquenazi explains using the Simondonian text. He does not recognise a psychological philosophy in *IPC* because he does not find the centre in the theory of individuation in consciousness or in the subject, but this does not prevent the sociologist from exploring a still little-trodden ground, that of the theory of emotion [6, p. 51]. Esquenazi does not have access to any critical literature on Simondon's theory from a psychological perspective, and the rediscovery of Simondon's work was still in its infancy. (the *Cahiers Simondon* edited by Jean-Hugues Barthélémy did not see the light until [31]. The sociologist, however, identifies as the nodal point of Simondon's work the notion of *orientation* that allows one to bypass the primacy of objectivity or subjectivity [6, p. 51]. What is primary in Simondon is the relationship between two individuals or, primarily, between an individual and the environment that orientate each other, that is, they take meaning from each other. This orientation occurs by displacement, i.e. by an increase in tension between the two individuals or between the individual and the environment after a balance has been found.

Out of this consideration, Esquenazi finds in Simondon a process of emotion divided into three essential phases. The first stage considers affect as distinct from emotion. Affection, for Simondon, concerns the individual and occurs when there is an encounter with another individual that generates tension causing the individual to be totally immersed in that specific affection (joy, sorrow, despondency, etc.). In this first phase framed by Esquenazi the affection remains only in the individual and does not enter the process of individuation, that is, it does not establish a relationship with the other, with the environment. In these lines Esquenazi speaks of individuation, but if we were to stick to the Simondonian lexicon, we would have to say that affection does not enter into the process of transindividuation [6, p. 52]. The second stage noted by Esquenazi is that in which affection is transformed into emotion through the process of transindividuation, i.e. through the «participation au milieu» that enables the individual to situate the affection. To do this, the individual must abandon his singularity and enter the collective, the only space in which signification occurs. Affection becomes emotion and as such can prolong and propagate itself thanks to the possibility of detaching itself from the individual or the environment that provoked it, thus becoming, says Esquenazi, «the site of a reaction that organises exchanges in the environment but also the restructuring of the latter» [6, p. 52]. Emotion can only become such if it is shared by the collective, i.e. if it detaches itself from the individual by becoming transindividualised. Finally, the third phase identified by the sociologist is that of the transfer of emotion into language, in the sense of a passage towards the condition of mediator and creator of meanings. In this sense, Esquenazi concludes his reconstructive work by affirming that in Simondon, the transcendental subject, i.e. the collective, is what each individual must relate to and, therefore, situate himself in at the moment in which an emotion emerges from it, i.e. the request for a new balance and a new language. It is thus emotion that opens up the possibility of situating oneself in a collective, just as affect is what allows the individual to situate himself in the world, to break through the two-dimensionality of perception to find the third dimension, that of depth.

4.5 Jean-Hugues Barthélémy and Anxiety Between Simondon and Heidegger

Before editing *Cahiers Simondon* [31–36], Jean-Hugues Barthélémy devoted four monographs to Simondonian thought [2, 37–39], but only in the last one, the 2008 one in which he describes Simondon's philosophy as 'genetic encyclopaedonism', does he carve out a small space to analyse an important aspect of Simondonian psychology. Barthélémy also does not directly enter the order of consideration of a genuine Simondonian psychological theory, only pursuing psychological themes in relation to a more general theory of the trans-individual and the anthropological question [2, p. 87]. Barthélémy focuses on the third part of *ILFI*, the part on psychic individuation, starting with the complex thesis on anxiety. Looking again at this part of *ILFI*, it is possible to note that the paragraph devoted to anxiety is found only halfway through and, therefore, certainly linked to what Simondon expounds in the preceding pages. Barthélémy notes how analysing anxiety in *ILFI* also means entering into the question of the transindividual, a term that Simondon takes from Raymond Ruyer but completely transforms in meaning [2, pp. 87–88]. The passage through anxiety would turn out to be necessary for access to the transindividual just as the psychic would constitute a transitory route towards the collective. Now, to dismiss the psychic as a transitory pathway within the individual that brings a problem that can only be solved in the collective seems to be an excessive reduction with respect to the Simondonian dictate. This is not the case with the analysis carried out on anxiety and its relationship with emotion. Barthélémy makes no mention of the theory of emotion, limiting himself to inserting the Simondonian definition in the comparison with anxiety. Even from Barthélémy's analysis, however, it emerges how emotion is what allows access to the trans-individual without stopping at the inter-individual, a condition for which no new individuation is needed, but only a structural analogy between the two interacting individuals. What Simondon seeks, however, is not a structural analogy, but a functional analogy, which means that the interindividual remains on the threshold of the collective without being able to enter it. Instead, emotion is the *conditio sine qua non* of entry into the collective through transindividuation. This is possible because, as Barthélémy points out quoting *ILFI*, emotion is the «the capacity of the individuated being to de-individuate temporarily in order to participate in a wider individuation» [2, p. 89]. Through emotion, the living individual loses his momentary stability and de-individualises himself, i.e. he is displaced from the solution he had found because a new problem is placed before him that cannot be solved with what he has at his disposal as a single individual.

Defining emotion as a condition for transition into the collective through its de-individuating power seems to hand over its power to a specific and apparently negative emotion such as anxiety, when it is stated that «is like the reverse journey of ontogenesis; it unravels what has been woven, it goes backwards into all the senses» [2, p. 89]. So, it is anxiety that intertwines the webs woven by the different stages of individuation that Simondon described, de-individualising the single individual. In this sense Barthélémy argues that the difference between the individuating capacity of emotion in general and the specific function of anxiety is subtle and seems to reside in the provisional character of de-individuation itself. Not to mention the fact that, thought of as a vector of trans-individuation, the very proximity of anxiety to emotion should give the former a positive potential. Barthélémy notes, however, how Simondon appears to privilege the negative aspect of anxiety, that is, its being a complete dissolution of being in the attempt to solve the problem posed by de-individuation without the transition to the collective.

From this analysis of anxiety in Simondon, Barthélémy concludes the argument by tracing a link between the French philosopher and Heidegger:

> The author wanted to think of it in the way Heidegger had characterised *Dasein*, but this time by anchoring it in the affectivity with which living animals are endowed, and by making anguish that very special emotion which opens the way to the realisation, very probably impossible, of the I without the We, or of *personality* without *transindividuality* [2, p. 90].

Simondon would thus rethink the Heideggerian being-ness as something that belongs to all living animals and is dominated by affectivity. In this context, anxiety would be the only possible path to the realisation of the solipsistic ego, i.e. a personality detached from the collective. Barthélémy argues that this characterisation of anxiety by Simondon was determined by the need to explain an experience that the philosopher himself lived. Anxiety is characterised, therefore, as a mental pathology that appears when the pre-individual charge united with the single individual is unable to meet another pre-individual charge with its relative individual. From this condition of impossibility of trans-individualisation, mental pathology arises, while anxiety seems to be its cause.

Barthélémy emphasises that the strong emphasis on trans-individuality in Simondon does not imply that the author dissents from the condition of solitude. On the contrary, Simondon states that trans-individuation can only be achieved if one passes through the «eprouve» of loneliness [2, p. 91], i.e. that phase of the individual's doubling in which he questions himself. That between the individual and his double is the first transindividual relationship that is established. In this presence of transindividuation already within the single individual, Barthélémy finds a paradox, for in this way it would seem as if Simondon could no longer clearly define what the transindividual is and at what stage it can be traced [2, p. 93].

As a matter of fact, the notion of the transindividual emerges in *ILFI* as early as the end of the second part and is then deepened in the third part. This could demonstrate how transindividuation is not a process specific to only one phase of being, that which passes from the psychic to the collective, but a movement that somehow remains internal and external to the individual in each of its phases. The transition from the psychic to the collective would thus be, so to speak, the exploit of this general movement.

4.6 The Italian Reception: Between Paolo Virno and Giovanni Carrozzini

The first work in Italian on Simondonian psychology appeared in 2008 and concerns precisely the theory of affectivity. It is included in the twelfth volume of the journal *Il Protagora* dedicated to Gilbert Simondon's thought and edited by Giovanni Carrozzini. It is a very important volume for the Italian reception of the French philosopher's thought since before it the only reference in the context of Italian studies was a text by Paolo Virno dedicated to the sociopolitical implications of Simondon's theory of individuation [40].[3] Also in 2010, Andrea Bardin's first work appeared, like Virno's, dedicated more to the political aspect of Simondonian philosophy [41].

The article in *Il Protagora*, by Luca Pinzolo, attempts to trace the points of contact and distance between phenomenologist Michel Henry's theory of affection and Simondon's theory of affectivity [42, pp. 427–433]. From the very first lines, Pinzolo emphasises that this juxtaposition must come to terms with the Husserlian phenomenological tradition in which Henry's thought is embedded and, therefore, with the concept of intersubjectivity, which seems very distant from that of transindividuality used by Simondon [42, p. 427].

Pinzolo states at the outset that he understands the concept of metastability used by Simondon differently: «a denotative concept of an individuation that is produced from something that is less than an individual and at the same time more than an individual» [42, p. 427]. However, for the French philosopher, metastability is understood in general as the condition that allows for individuation in the sense of a never completely stable equilibrium that generates problems to which the structuring of an individual is the momentary solution. Pinzolo states that his attempt is to demonstrate that this 'something' that is less and more than the individual resides in the affects and he dedicates the last part of the article precisely to the Simondonian theory

[3] Available at the following link: www.filosofia.it

of affectivity, arguing that, unlike phenomenology, the Saint-Etienne philosopher's perspective does not refer to a single individual but to the process of individuation and that, therefore, the study of the other and the relationship with it is markedly different from phenomenological thought. Affectivity itself is not linked to a self, an *ego*, but to «collective representation and actions» [42, p. 432]. Pinzolo emphasises how, in Simondon, the collective is not configured as an encounter of atomised individuals (in Simondon's terms: not as interindividuality) but as an «event of individuation that includes relationality in itself» and that is only discovered through affection/emotion [42, p. 432]. Affection and emotion thus turn out to be the place of exchange between the pre-individual and the individuated from which a problem is generated that is impossible to solve in the single individual. In its representation of a very early opening up of Italian literature with respect to Simondon's psychological reflection, Pinzolo's essay does not yet take into account, among other things, the fundamental difference between affection and emotion present in *ILFI*. Moreover, the analysis does not go directly into Simondon's theory, mentioning it only in order to show the assonance with Henry's conception of affectivity in the reference (of common Merleau-Pontyan matrix) to the pre-individual charge.

In 2011 Giovanni Carrozzini devoted a substantial part of his monograph on Simondon to general psychology courses taught in the 1960s [42, pp. 101–164]. Significantly, it is in Italy that attention is brought to these texts, when even in France interest was struggling to take off. The *Cahiers* themselves arrive in 2011 at their third volume without reference to the courses of the 1960s. It must be said that at this date only the courses on perception had still been published [43]. In this, too, Carrozzini's archival work is very important in illuminating Simondon's psychology and the role it played in his overall project:

> The specific interest of this letter—as well as those that will follow [...] springs [...] from Simondon's, albeit implicit, enunciation of a precise scientific and philosophical orientation, which also underlies his experiments in the field of psychophysiology and psychology tout court. In fact, as has already been mentioned, Simondon was already developing an organic vision of reality during his high school teaching [...] [42, p. 106].

The letter to which Carrozzini refers is the one found in the archives of the Bibliothéque de psychologie Henri Piéron of the Université René Descartes Paris V and sent by Simondon in June 1950 to the experimental psychologist to express his interest in pursuing studies in psychology. In it, Simondon appears to emphasise that psychological research should not be separated from analysis and in-depth study of techniques. Considering this need to hold psychology and technique together, it would already seem evident how much the study of psychological processes played a founding role in the constitution of Simondonian philosophy.

Through the analysis of texts by Maurice Mouilland [42, pp. 401–404] and Emile Jalley (2007), Carrozzini reconstructs the path followed by Simondon during his teaching of psychology first at the University of Poitiers and then at the Sorbonne in the 1960s and 1970s. Those spent at Poitiers were certainly more serene years than his Parisian ones, albeit punctuated by 'stormy' relations with a colleague who disapproved of the prospect of a study of psychology open to other disciplines.

The Parisian period, on the other hand, was marked by an ongoing struggle with the group of psychologists who wanted France to follow the American direction, veering towards a psychology applied to the industrial work. This push, brought about by Maurice Reuchlin who directed the differential psychology laboratory, would lead, around 1976, the University Council to take the decision to replace Simondon with Reuchlin himself in the direction of the general psychology course.

This reconstruction allows Carrozzini to state how there was a real 'ostracism' towards Simondon that would have hindered his career in psychology; and this because of his philosophical training and his encyclopaedic spirit [4, p. 115]. Simondon's presence among the ranks of psychology professors first in Poitiers and then in Paris would have undermined the desire to distance psychology from philosophy in order to give it a scientific status. To substantiate this thesis, Carrozzini cites one of the first texts in which Simondon set out his position on the status of psychology: the *Histoire de la psychologie moderne* written in four hands with François Le Terrier and published in the *Encyclopédie la Pléiade* in 1957.

An original feature of Carrozzini's interpretation is both the underlining of the philosophical root that Simondon traces in psychology, and the attention in determining a possible closeness between the task that Simondon assigns to psychology and that of philosophy. Thus, even if primacy appears to be assigned to philosophy, both disciplines seem to have the task of unifying or, rather, re-unifying the sciences. This only emphasises the Simondonian inclination towards the encyclopaedic spirit, i.e. towards the need to rediscover a «sense of the whole» through a return to the stage when divisions and specialisations did not exist [4, p. 117].

Among the issues noted by Carrozzini is that of the link established by Simondon between the study of the human sciences as a whole (hence also of psychology) and that of techniques. It has been emphasised how in the workshop directed at Poitiers and then in the one set up at the Sorbonne the theoretical study of psychology was always accompanied by knowledge and application of techniques [4, p. 120]. According to Carrozzini, it was precisely this attention to technicality and the operative relationship between organisms and the environment that made Simondon's perspective on psychology innovative and different from those circulating in those years [4, p. 126]. There are two fundamental junctions of this fruitful intertwining: the finding in the technical field of paradigms and schemes to reconsider the relationship between organisms and the environment and the use of the concept of modulating amplification as a «methodological principle» to follow the course of psychic activity [4, p. 126]. This second junction is traced by Carrozzini in particular in the articulation of the courses Simondon held at the Sorbonne in the 1960s. It is important to emphasise that until the translation of the collected work *Sur la technique*, published in Italy only in 2017, and the course *Imagination et invention* whose Italian translation is from 2023, there was no access to these texts except in the original language.

Carrozzini emphasises how Simondon was able to establish the connection between psychology and technique starting from a rethinking of the method of analogy as a tool capable of connecting no longer structures, but operations. In this sense, technique provides the «interpretative paradigms», while psychology, if connected through functional/operational analogy with technique, turns out to be the field of application of the paradigms themselves [4, pp. 128–129]. Therefore, it is thanks to the functional analogy that Simondon is able to reread the history of psychology in function of a reunification of the different schools of thought in the direction of a science of man in which all the human sciences would converge.

Carrozzini's aim is to highlight the «mentalité technique» present not only in the works strictly related to technique and in *ILFI*, but also in the psychology courses Simondon taught during his time at the Sorbonne's UER. Carrozzini points out how this mentality is not to be understood in a manipulative sense, but in the original sense of an operative participation on the part of the living organism in an operativity intrinsic to the being itself. This primacy of operativity can also be highlighted by the pre-eminence that Simondon assigns to motricity over sentience [4, pp. 130–132]. Every living organism first moves in order to orient itself in the world and only secondarily perceives something. This role assigned to motricity, together with the use of technological paradigms to explain vital processes, could have placed Simondonian thought in a mechanistic tradition. Carrozzini, on the other hand, highlights the clear reference to the work of Georges Canguilhem, following which Simondon's theory would fit into the path of a possible reconciliation between vitalist and mechanist positions [4, note n. 49, 57, pp. 134–139]. Simondon's use of technological paradigms does not neutralise the aleatory and indeterminable character of the vital but serves to demonstrate that every process of organisation is operational and, as such, cannot be completely automatic. In the *dis-automatization* of operational processes, there is the impossibility of reducing Simondonian thought to mechanicism and determinism and the opening up to an organism-machine relationship of analogy that is no longer structural but functional, as Canguilhem himself hoped [4, note n. 57, pp. 134–139, 44, p. 130].

Carrozzini continues the analysis by following the chronological order of Simondon's courses with the aim of bringing out the amplifying aspect of psychic processes. In this perspective, his work is certainly pioneering and makes it possible to come into contact with the texts of Simondon's psychology courses that had remained largely unknown up to this point, opening the way also to the possibility of recognising a specific Simondonian psychology and interpreting it as one of the pillars of the whole.

4.7 The Anglo-Saxon Reception: Tucker's *Affective Technological Consciousness* and Celis Bueno and Schettini's *Transindividual Affect*

4.7.1 A. Tucker's Affective Technological Consciousness

In English, interest in Simondonian psychology comes in particular from Ian Tucker, lecturer in social and health psychology at the University of East London, who began using Simondon's theories in [7–9]. In the first article dedicated to these topics,

Tucker presents his interest in Simondonian theses starting from an exquisitely psycho-social problem: surveillance in the relationship between bodies and technology. The aim of Tucker's work is to rethink information as a mediator in the relationship that bodies have with technologies and to analyse the relationship between what he calls «affective technological consciousness» [7, p. 30]. Simondon's theory is in this sense intended to support the author's original argument in determining the impossibility of thinking about and improving affective technological consciousness without referring to embodied consciousness. In fact, Tucker argues that in the research carried out on surveillance, much emphasis has been placed on the dehumanising aspect of the technologies that perform this role, but no emphasis has been placed on how much human there is in the body-technology relationship.

Instead, the Simondonian perspective on information as a creative force and the idea that technologies are not mere tools for human beings, but individuals in their own right, capable of developing and expressing creativity, are of fundamental support here. As Tucker points out, Simondon's view of the individual as process rather than object/subject allows us to think that every time a human individual interacts with technology, he is already subject to information, i.e. he changes his organisation to find a new metastable equilibrium with the technical object. In Tucker's words, subjects are «fleshy» forms of information [7, p. 34].

The crux of this use of Simondonian theories is the fact that for the philosopher from Saint-Étienne, precisely because they are processual, individuals who are already always informed identify themselves from a pre-individual, and this happens through affectivity.

Tucker emphasises this because he believes that this stance of Simondon's can be a fundamental reflexive junction (in his view, of Spinozian origin) that can open up a new study of surveillance in the body-technology relationship. This is because the affective drive is guided by the information that organises and structures it. In other words, affection passes through the individual body, organising its relationship with the other and thus flowing into the trans-individual and the collective. To this Tucker adds:

> The notion of affect keeps in the analytic spectrum the *human* element of technologically-mediated surveillance systems, as focus shifts from the technical capacities for surveillance in modern information technologies to the forms of embodied experience made possible therein. [...] Affect as a concept also directs us towards the *intensive* elements of experience in terms of the collective ineffable and yet *felt* form of bodily experience in relation to technologies [7, pp. 38–39].

The insistence on the Simondonian conception of affect allows Tucker to rethink technological surveillance not only as a technique aimed at control, but as an experience in which the human element is prevalent and opens up new possible identifications. Surveillance is thus not an external gaze on human individuals and their living space but a mode of expression of information processes as transformative and creative. People are not surveilled by being invested in the informational process, but respond affectively and emotionally to surveillance by constructing their own affective-emotional sphere. This construction is non-conscious as affectivity operates from the pre-individual layer which, as Simondon argued, works on the unconscious.

Tucker proposes to think that information flows also present in surveillance move in the pre-individual and as such cannot be captured and mapped clearly, but can be felt [7, pp. 40–41],in this sense recalling the concept of affect as a force that pushes or holds the action of individuals. This same force that comes from information flows is also present in human individuals using those technologies that seem to monitor them. In fact, Tucker concludes that power is, therefore, neither only on the side of technologies nor on the side of human individuals since on both sides there is the presence of the human element given by the creativity of information.

The 2018 article goes even further into the merits of the Simondonian psychological question through the possible intersection with Spinoza and Deleuze. Tucker argues that through their thinking about affectivity in becoming it is possible to trace «potentially valuable lessons» both for psychology in general and for contemporary challenges in relation to the interaction between society and technology [8, p. 129]. In particular, the attention that these three philosophers gave to affectivity could be decisive with respect to the contemporary thesis that sees affectivity/emotionality as the first and main human area to be attacked and threatened by machines.

Starting with the Deleuzian Spinoza, Tucker argues that the analysis on affects carried out by the Dutch philosopher is fundamental to an extended approach to psychology that would no longer position psychological processes solely within individuals. In this direction, the theory of *distributed cognition* argued as early as 1995 with Edwin Hutchins that psychological processes transcend the physical limit of the individual human brain and can also be extended to computers, which are just as much agents as other individuals (Hutchins 1995). However, Hutchins' is still a computational approach, whereas Tucker's aim is to find a theory that extends psychological processes without being computational. The concept of affect found in Spinoza and taken up by Deleuze could be the key to tracing this new extended psychology. Indeed, in the Deleuzian thesis, psychological processes occur on a «continuum» that goes beyond the brain, but also beyond human and non-human bodies [8, p. 131]. This *continuum* is fundamentally, and before anything else, affective and, therefore, determining the degree or potentiality of action of a body immersed in it.

Given the Spinozian derivation of the concept of affect present in Deleuze, Tucker points out how much the whole of Deleuzian thought was also influenced by Simondon and his attempt to situate psychological processes in the dynamics of individuation. This implies, if one rereads *ILFI* entirely and carefully, that psychological analysis cannot disregard taking into account the *milieu associé* with which the individual stands in relation. Psychological analysis is thus always the analysis of a relationship [8, p. 134]. Affectivity comes into play precisely in this relationship between the individual and the *milieu associé* since it can be identified with that feeling of never being completely alone. In Simondon, affectivity is in this sense what allows the individual to recognise himself as always transindividual, i.e. capable of transcending the boundaries of the body as only a metastable state of a *continuum.* This implies that affectivity «is consequently the fundamental inter-connectedness of individual and pre-individual in the formation of psychic being» and, consequently, is the force that moves psychic processes [8, p. 138].

Thus, Tucker recovers the concept of affectivity found in Simondon, explains its Spinozian derivation and the influence it had on Deleuze, and aims to demonstrate the importance of this transindividualising power of the affections with respect to the relationship with the digital world. A world, the latter, with which Simondon never dealt but on which, instead, his psychological theories would still have much to say. As Tucker argues, despite the fact that the general concern is still directed today, as it was at the time of *MEOT*, towards the idea that digital technologies can eliminate *realitè humaine* (*MEOT* [47], p. 15) and replace the human, Simondon's theory of psychology from an ontogenetic perspective makes it possible to affirm that this threat cannot be realised. This is because:

> It conceptualises psyche and collective as always-already intertwined, and in doing so, widens the unit of analysis to consider the conditions in and through which individual and collective life unfold. In relation to living in increasingly digital environments, a focus on individuation encourages awareness of how we are shaped by phases of metastable being, and ever subject to individuation by future collectives [8, p. 140].

This approach, according to Tucker, would make it possible to look at the relationship with the digital no longer as threatening, but as creative and collective. Such a change of perspective would also be very useful for artificial intelligence studies, which have focused on individualist ideas of the psychological world of affectivity and emotion. Instead, if one considers, as Simondon did, the psychological world as part of an ontogenetic process in which the individual is never isolated from his or her *milieu associé* and other individuals, then more accurate studies can be made of the capacity of digital technologies or machines to be affected or emotional. With this in mind, Tucker concludes by stating that only if we rethink the human way of being affected and emotionalised as always trans-individual can we understand how to study psychic life in the digital age [8, p. 141].

In a more recent article in 2021, Tucker takes up the discussion opened in 2018 by further articulating Simondon's theory of emotion [9]. In particular, from an analysis of individuation theory, Tucker addresses the psychological question by emphasising its being part of this process and its emergence in the face of a problematic relationship that arises when the individual begins to orientate himself in the environment. Tucker emphasises that when Simondon speaks of perception as the moment when the individual begins to orientate himself in the world, he does not mean that previously the individual and the environment were separate. The individual emerges from the pre-individual always together with its associated environment and, as such, is first and foremost a relation. Perception is the problematic moment when the individual becomes aware of this associated environment and begins to come to terms with it. Initially as something that seems distant from him, objectifying it, and later, thanks to affectivity and emotion, realising that it is part of him, of *that always us*. Thus affectivity comes into play when the individual becomes aware of something outside himself and must begin to come to terms with the pre-individual drive that this something conveys:

> While perception acts to resolve the tension between the body and environment, affectivity is the «deeper» relation of individuation and pre-individuation. The tension that affectivity exists to resolve is «closer» to the individual as «subject being» [9, p. 7].

Affectivity is thus the relationship between the individual being individuated and the pre-individual that has remained bound to him from the moment of genesis. In this sense, Tucker can say that the resolution at which affectivity works brings the individual closer to his subject being because for Simondon the subject is that phase of being in which pre-individual, individual and trans-individual manage to find a metastable equilibrium. Now, affectivity, like perception, must solve a problem of incompatibility (between individual and environment or between individual and pre-individual) not so much because it is a relationship that is established between different materials, but because it has the problem of coupling different modes of being in a functional nexus. Hence, as the need to find a solution that embeds them in a new play of forces without reducing them to each other.

If perception and affectivity are resolving processes of different modes of being and are part of what Simondon calls psychic individuation, then, Tucker reminds us, psychic life cannot be considered as only internal to the individual. Psychic life is necessarily relational like the entire process of individuation: «it exists psychically as a relational being that must operate in and through orientations with itself as an individual and as part of a collective (milieu)» [9, p. 7].

Taking into consideration Simondon's general theory of individuation and focusing on the importance of affect and emotion in that phase of the process called psychic individuation, Tucker devotes the last part of his work to demonstrating that:

> Digital practices have come to increasingly feature in the operation of individuation—as they constitute large parts of the relations we have with ourselves (as subject) and others (as collective)—rendering them in direct dialogue with emotion and affectivity [9, p. 11].

Tucker's problem is that in the conception of artificial intelligences that can read emotions, the model referred to is not Simondon's relational model, but the one according to which emotion can be revealed through physiological traits because it is something individual. The Simondonian approach, on the other hand, would make it possible to define the emotional relationship between individual and algorithm as an individuation and, therefore, without there being a prevarication of one party over the other. There is, therefore, not an algorithm that reads the emotions of the individual, but an algorithm that takes part in the psychic structuring of what may be called the subject. Algorithms and the machines that contain them are no longer tools for collecting and processing data, but agents of the psychological structuring process. They become, in fact, part of *the always us that* everyone is. Only from this revision, Tucker argues, is it possible to talk about psychological life in digitally mediated environments and to re-evaluate data, so jealously guarded, no longer as individual, but always as collective.

Sticking with what at this point seems to be a fruitful interaction between Simondonian psychology and data sciences, in 2023 Tucker insists on the importance of Simondon's reconceptualisation of information and emotion in understanding the interaction between individuals and data. In particular, in this last article Tucker draws attention to an issue that is often overlooked, as it is more thematised by his follower Stiegler: the shift from thinking of individuation as spatialised to a temporalised conception that leaves room for potentiality (Tucker 2023, p. 9). Tucker argues that one should begin to think of data collected by contemporary technologies as temporal processes that operate alongside processes of individuation. These can be termed «emotional in-formation» for Tucker (2023, p. 9), emotional in-formations that proceed together with psychic individuation. In order to explain the notion of emotional in-formations, Tucker gives the example of a post that is published on an online forum and, as soon as it is put online, passes into the forum's collective. Whoever comments on that post enters the timelines of the forum, of a collective, and not of the individual who wrote it. In this sense, therefore, Tucker concludes, data are «in-formational, future-oriented and unspecified-in-advance forces of transformative potential» (Tucker 2023, p. 12) that may seem to refer to individuals as initially the process of psychic individuation seemed to be of the individual but, in truth, once generated, are no longer enclosable in representations of the individual. Any data that is generated is in fact already always transindividual (Tucker 2023, p. 12).

4.7.2 B. Celis Bueno and Schettini's Transindividual Affect

A proposed reading of Simondon's theory of emotion in a posthuman key comes in 2022 from Celis Bueno and Schettini. It is a perspective that intends to redefine the human from a non-anthropocentric and non-dualist point of view through the decentralisation of rationality as a privileged human belonging. Questioning affectivity and emotionality from a relational ontology such as that of Simondon seems to be fundamental for the two authors [5].

Celis Bueno and Schettini argue that for Simondon the case of affectivity and emotion is a fundamental key to explaining the process that they define as transductive. The two authors place Simondon's philosophy in a line that leads to what is defined as the contemporary «affective turn» and which occurred from the idea of directing the study of the human no longer on rationality, but on the body, the sensorial and the dynamic [5] (Paasonen, Hills and Petit 2015). In particular, these authors propose a posthuman theory of emotion that is based on four principles transferred from Simondonian philosophy: questioning the pre-eminence of rationality over emotionality, thinking of emotion as relational and not tied to the individual, taking into account the non-representational dimensions of subjectivity and, finally, linking emotion and affect also to non-human individualities. The key concept they coined from Simondon's thesis is that of «transindividual affect» [5, p. 124]. In *ILFI* it emerges how much the problematic that arises between affect and emotion cannot be resolved except through trans-individuation and the transition to the collective. Affections, as the pathway through which the living individual relates to the pre-individual, orientate the individual with respect to the other from himself, which, once encountered outside himself, resonates within him.

In this sense, then, Simondon seems to be proposing a psychological conception that cannot be dissociated from the social one (a theme widely debated in the 1960 report) since psychic individuation and the collective move on the same terrain, that of trans-individuality. Moreover, the latter is the same space in which affectivity and emotionality move as neither purely individual nor purely socio-cultural expression. Finally, thanks to Simondon's affect-emotion theory, it is possible to think of a *disindividuation* of the individual every time an affect, and then an emotion, are generated, since through these they participate in a broader individuation, that of the collective.

These recent works of analysis in English have certainly shed some light on Simondon's theory of emotion and affectivity and find in it a fundamental junction for crucial contemporary issues. From a historiographical as well as a reconstructive point of view, however, it is evident that there is still a lack of systematic study of the courses and publications following the '58 theses. The primary cause of this is the late publication and lack of translation into other languages of the collections. However, one cannot fail to note how greater attention to Simondon's entire course would open up further discussions regarding both the question of the interaction between humans and technology from an emotional point of view and the posthumanist critique with which Simondon has points of contact but also several points of possible contrast.

References

1. Simondon G (2014) Sur la technique, 1953–1983. Puf
2. Barthélémy JH (2008) Simondon ou l'encyclopédisme génétique. Puf
3. Van Caneghem D (1989) Hommage à Gilbert Simondon. Bulletin de Psychologie 42(392):815–836
4. Carrozzini G (2011) Gilbert Simondon filosofo della "mentalité technique". Mimesis/Centro Internazionale Insubrico
5. Celis BC, Schettini C (2022) Transindividual affect: Gilbert Simondon's contribution to a posthumanist theory of emotions. Emot Rev 14(2):121–131
6. Esquenazi JP (2004) Vers la citoyenneté: l'étape de l'émotion, *Mots*. Les langages du politique 75:47–57
7. Tucker IM (2013) Bodies and surveillance: simondon, information and affect. Distinktion: J Soc Theory 14/1:30–41
8. Tucker IM (2018) Deleuze, Simondon and the 'Problem' of Psychological Life Maria Nichterlein, Putting the Deleuzian Machine to Work in Psychology. Annu Rev Crit Psychol 14:127–143
9. Tucker IM (2021) Simondon, emotion, and individuation: the tensions of psychological life in digital worlds. Theory Psychol 32(1). https://doi.org/10.1177/09593543211055199
10. Tucker IM (2024) The emotional in-formation of digital life: Simondon, individuation and affectivity. Distinktion: J Soc Theory 25(3):282–296
11. Stiegler B (1994b) La technique et le temps, I-La faute d'Epiméthée. Galilée
12. Stiegler B (1996) La technique et le temps, II-La désorentation. Galilée
13. Stiegler B (2001) La technique et le temps, III-Le Temps du cinéma et la Question du mal-être. Galilée
14. Lagache D (2013) L'unite de la psychologie: psychologie expérimentale et psychologie clinique. Puf
15. Tenti G (2024) Tecnogenesi della Terra Simondon e il neo-magico. Itinera 27:94–110
16. Combes M (1999) Simondon, Individu et collectivité: Pour une philosophie du transindividuel. Puf

17. Stiegler B (1998) Temps et individuations technique, psychique et collective dans l'œuvre de Simondon. Intellectica 26–27:241–256
18. Stiegler B (1992) Leroi-Gourhan, part maudite de l'anthropologie. Les nouvelles de l'archéologie 48–49:23–30
19. Stiegler B (1992) La programmatologie de Leroi-Gourhan. Les nouvelles de l'archéologie 48–49:31–36
20. Chabot P (2016) ChatBot le robot: drame philosophique en quatre questions et cinq actes. Puf
21. Hui Y (2016) On the existence of digital objects. University of Minnesota Press
22. Chabot P (2003) La philosophie de Simondon. Vrin
23. Jung CG (1966) The collected works of C. G. Jung. Princeton University Press
24. Esquenazi JP (1994) Film, perception et mémoire. L'Harmattan
25. Esquenazi JP (1999) Télévision et démocratie. Le politique à la télévision française, 1958–1990. Puf
26. Esquenazi JP (2009a) La vérité de la fiction. Comment peut-on croire que les récits de fiction nous parlent sérieusement de la réalité? Hermès Science Lavoisier
27. Esquenazi JP (2009b) Mythologie des séries télé. Le Cavalier Bleu Éditions
28. Esquenazi JP (2010) Les Séries Télévisées. L'avenir du cinéma. Armand Colin Cinéma
29. Sunseri F (2025) L'umanesimo tecnologico di Gilbert Simondon. Giornale Critico di Storia delle Idee 1:165–178
30. Deleuze G (1968) Différence et répétition. Puf
31. Barthélémy JH (ed) (2009) Cahiers Simondon, vol 1. L'Harmattan
32. Barthélémy JH (ed) (2010) Cahiers Simondon, vol 2. L'Harmattan
33. Barthélémy JH (ed) (2011) Cahiers Simondon, vol 3. L'Harmattan
34. Barthélémy JH (ed) (2012) Cahiers Simondon, vol 4. L'Harmattan
35. Barthélémy JH (ed) (2013) Cahiers Simondon, vol 5. L'Harmattan
36. Barthélémy JH (ed) (2015) Cahiers Simondon, vol 6. L'Harmattan
37. Barthélémy JH (2005a) Penser l'individuation. Simondon et la philosophie de la nature. L'Harmattan
38. Barthélémy JH (2005b) Penser la connaissance et la technique après Simondon. L'Harmattan
39. Barthélémy JH (2014) Simondon. Les Belles Lettres
40. Virno P (2001) Moltitudine e principio di individuazione. Filosofia.it. www.filosofia.it
41. Bardin A (2010) Epistemologia e politica in Gilbert Simondon. Individuazione, tecnica e sistemi sociali. FuoriRegistro
42. Carrozzini G (eds) (2008) Gilbert Simondon filosofo delle tecniche. Il Protagora 36/12. Mimesis
43. Simondon G (2006) Cours sur la perception (1964–1965). La Transparence
44. Canguilhem G (2015) La connaissance de la vie. Vrin
45. Simondon G (1989) L'Individuation psychique et collective. Aubier [IPC]
46. Simondon G (1995) L'Individu et sa genèse physico-biologique. Millon, Grenoble [IGP]
47. Simondon G (2017) On the mode of existence of technical objects. Univocal Publishing [MEOT]
48. Simondon G (2020) Individuation in Light of Notions of Form and Information. University of Minnesota Press [ILFI]
49. Stiegler B (1994a) La maïeutique de l'objet comme organisation de l'inorganique. Gilbert Simondon: une pensée de l'individuation et de la technique. Albin Michel

Chapter 5
Conclusions

The research conducted in this book confirms and expands upon the thesis put forward in the *Introduction*: It is evident that Gilbert Simondon's interest in psychology was not a fleeting interest or a consequence of his early education; rather, it was a fundamental element of his entire intellectual enterprise. Psychology is not confined to a specific disciplinary field; rather, it is characterised by a transversal, constitutive value in Simondon. This capacity enables it to articulate ontological, technical, vital and relational levels in a single theory of individuation.

From his earliest writings on cybernetics, Simondon questioned the reductionist views of humanity, opposing any attempt to assimilate the psyche to a simple computer or to think of the subject as a passive container of representations. The proposal under discussion is oriented towards a genetic psychology based on three principles: metastability, creative discontinuity of individuation and temporal irreversibility of psychic processes. As evidenced by his pedagogical articles from 1953–1954, the notion of a psychology that does not dissociate subjective from relational data can be traced back to this period. Instead, the objective was to comprehend the dynamics through which an individual constitutes themselves as such – not in isolation, but always within a transindividual tension. It is on this basis that, in the course of 1956, a radical reformulation of the foundations of modern psychology takes shape.

This theoretical path involves a comprehensive rethinking of the structure of Simondonian knowledge. The two doctoral theses, initially conceived as a philosophical study of psychological experience, bifurcate in two directions: on the one hand, an ontology of relation (*ILFI*); on the other, a philosophy of technology (*MEOT*). Nevertheless, the present volume has sought to demonstrate that this bifurcation should not be understood as a thematic dispersion, but rather as a process of ramification that finds its unifying principle in genetic psychology. Psychology thus becomes pivotal in the reconceptualisation of these two theses, and in the meticulous examination of the entire Simondonian construct as a endeavour to establish a general

F. Sunseri, *Psychology and Technology in Gilbert Simondon*, UNIPA Springer Series,
https://doi.org/10.1007/978-3-032-25740-6_5

allagmatics of being. In this theoretical framework, the psyche assumes a propulsive and transformative role.

This interpretation has been corroborated by comparison with later, little-explored texts. The essay *Formes, information, potentiels* (1960) demonstrates how psychic ontogenesis cannot be conceived outside of an informational and relational context, and how the subject itself is a structurally open and perpetual reality. The courses taught by Simondon in the 1960s and 1970s, in particular *Imagination et invention* (1965–66), confirm the philosopher's urgent need to conceptualise the psyche as a function of mediation between the individual and the world, but also as a creative power capable of transforming both. Imagination, as a metastable process of articulation between affectivity, knowledge and technique, thus emerges as one of the fulcrums of psychic experience – not as a mere faculty, but as an ontogenetic vector.

The present work aims to contribute to a comprehensive reconsideration of Simondonian genetic psychology. This reconsideration will not be considered from the perspective of one of the philosophical "uses" of his thought, but as the methodological and theoretical core of his project. In this sense, psychology is not merely an interpretative apparatus; it is the *conceptual laboratory* in which Simondon elaborates the conditions of possibility for thinking individuation as a relational and situated process, always poised between the technical, the living and the symbolic. In contrast to the prevailing view that subjectivity is merely a cognitive function, Simondon emphasises its affective and creative origins, positing these as the fundamental drivers of psychic and collective individuation.

The implications of this perspective are manifold. In principle, it proposes a reinterpretation of Simondon's texts in light of a semantics of affectivity and relation. From a historiographical perspective, it is necessary to expand the interpretative corpus to encompass scattered articles, untranslated lectures, notes, and materials that have hitherto been overlooked. The dispersal of psychological texts within collections not explicitly dedicated to the subject has been identified as a contributing factor to the marginalisation of this dimension in the international reception of Simondon's thought. A critical editorial project that recognises the unity of Simondon's psychological design – perhaps with new translations, commentaries and philological apparatus – would represent a fundamental contribution to the understanding of his work.

However, it is the epistemological and interdisciplinary implications that are arguably the most intriguing. Genetic psychology, in its Simondonian conception, now presents itself as a theoretical paradigm capable of dialoguing with neuroscience, complexity theory, educational studies, aesthetics and the philosophy of technology. It provides tools for understanding the processes of subjectivation within mediatised environments, for rethinking the role of creativity in digital contexts and for addressing the challenges of automation and psychic disaffection. In this sense, Simondon's approach can also be mobilised in a therapeutic key, as a form of philosophical care for the relationship between the individual, technology and the world.

From this standpoint, a fruitful avenue for further research emerges: a comparative analysis between Simondon's genetic psychology and Bernard Stiegler's concept of reflection. Commentators have previously noted the extent to which Stiegler has adopted and adapted the fundamental concepts of his mentor, particularly those pertaining to individuation, technology and memory. However, the present moment may signal a shift in perspective, necessitating a reinterpretation of Stiegler's work. Rather than perceiving him as a mere continuator, it may be more fruitful to consider him as a radical interpreter of Simondonian genetic psychology. In this light, his work can be regarded as having the capacity to elucidate aspects of the subject that have hitherto been implicit or underdeveloped. Stiegler's reflections on desire, affectivity, protension and the pharmacology of technique represent a further modulation of the philosophical project inaugurated by Simondon. The project is not content with explaining the functioning of the psyche; it aims to transform its conditions of possibility in the age of digital technologies.

In this sense, Simondonian genetic psychology can now be regarded as an open conceptual device: an organon for thinking about humanity beyond the classic oppositions between subject and object, nature and culture, individual and collective, technology and life. The objective is not merely to revive a theoretical project that has been overlooked, but rather to actualise its potential within a context in which questions concerning subjectivity, autonomy and technical relations are resurging with dramatic urgency. Genetic psychology, in its nascent phase, thus constitutes a philosophical and political resource for the present.

It can be argued that the primary significance of this research is in its demonstration that, for Simondon, psychology is not the domain in which pre-existing categories are applied; rather, it is the space in which new forms of thought, new images of the subject and new modes of coexistence are produced. This concept of a science of subjectivity can be regarded as a technique of living together.

MIX
Papier aus verantwortungsvollen Quellen
Paper from responsible sources
FSC® C105338

If you have any concerns about our products,
you can contact us on
ProductSafety@springernature.com

In case Publisher is established outside the EU,
the EU authorized representative is:
Springer Nature Customer Service Center GmbH
Europaplatz 3, 69115 Heidelberg, Germany

Printed by Libri Plureos GmbH
in Hamburg, Germany